Biosecurity for Livestock Diseases in Europe: Transmission, Risk Factors and Modelling

Biosecurity for Livestock Diseases in Europe: Transmission, Risk Factors and Modelling

Alistair Antonopoulos

Evelien Biebaut

Jeroen Dewulf

Johannes Charlier

Funded by the European Union

This work was funded by the European Union under the Horizon Europe grant 101083923 (BIOSECURE). Views and opinions expressed are, however, those of the authors only and do not necessarily reflect those of the European Union or the European Research Executive Agency (REA). Neither the European Union nor the granting authority can be held responsible for them.

CABI is a trading name of CAB International

CABI
Nosworthy Way
Wallingford
Oxfordshire OX10 8DE
UK

Tel: +44 (0)1491 832111
E-mail: info@cabi.org
Website: www.cabi.org

CABI
200 Portland Street
Boston
MA 02114
USA

T: +1 (617)682-9015
E-mail: cabi-nao@cabi.org

A catalogue record for this book is available from the British Library, London, UK.

ISBN-13: 9781800629813 (hardback)
9781800629820 (ePDF)
9781800629837 (ePub)

DOI: 10.1079/9781800629837.0000

Commissioning Editor: Alexandra Lainsbury
Editorial Assistant: Emma McCann
Production Editor: James Bishop

Typeset by Exeter Premedia Services Pvt Ltd, Chennai, India

Contents

About the Authors

Dr Alistair Antonopoulos

Alistair Antonopoulos is a scientific project manager working at Kreavet. He graduated from the University of Glasgow with a PhD in Veterinary Medicine. He previously developed novel molecular diagnostics for anthelmintic resistance and a next-generation sequencing assay to aid in the sustainability of livestock production. Alistair has been involved in multiple multidisciplinary and international projects. At Kreavet, he is contributing to the conduct and development of multi-actor Horizon EU-funded projects such as BIOSECURE, focused on improving biosecurity implementation within Europe for sustainable livestock production. As part of the BIOSECURE consortium, Alistair has been involved in the creation of a database of modelling parameters for livestock disease transmission, in addition to stakeholder disease prioritization exercises at the EU and participating country level.

Dr Evelien Biebaut

Evelien Biebaut is a postdoctoral researcher at the Veterinary Epidemiology Unit of Ghent University, currently serving as scientific manager for the Horizon Europe BIOSECURE project. She earned her PhD from the Faculty of Veterinary Medicine also at Ghent University, where her research focused on *Mycoplasma hyopneumoniae* infections in pigs, with a particular emphasis on transmission dynamics and immune responses. Passionate about advancing animal health and welfare, Evelien is committed to improving livestock management practices for the benefit of both animals and humans.

Professor Dr Jeroen Dewulf

Jeroen Dewulf serves as a full professor in veterinary epidemiology at the Faculty of Veterinary Medicine of Ghent University. His primary research interests include antimicrobial use and resistance in animals, as well as the prevention of epidemic and endemic diseases with an emphasis on biosecurity. He is passionate about turning data into veterinary expertise and policy for the benefit of animal and human health. Jeroen Dewulf is a prolific author with (co-)authorship of over 400 A1 publications in the field. Recently, he co-founded the spin-off company Biocheck.Gent and co-ordinates the EU research consortium BIOSECURE, which focuses on cutting-edge research in biosecurity in terrestrial animal production.

Dr Johannes Charlier

Johannes Charlier is a veterinary scientist and regularly invited speaker at international workshops or conferences. After a career of 13 years as veterinary parasitologist at Ghent University, and 2 years as scientific director at Avia-GIS, he founded the research agency Kreavet in 2017. He described his view on sustainable parasite and animal health management in a series of recognized review and opinion papers. The principles of those papers are a guideline for the development and strategies of Kreavet. Johannes is a EBVS® European Veterinary Specialist in Parasitology and has received several scientific awards for his contributions to veterinary parasitology and epidemiology. He is (technical) co-ordinator of European-funded projects including the secretariat of the STAR-IDAZ IRC, SPARC and DISCONTOOLS. He sits on the board of various scientific organizations and periodically acts as (guest) editor for scientific journals.

Foreword

Biosecurity in animal health refers to the set of measures implemented to prevent the introduction and spread of infectious diseases within animal populations. It involves various practices aimed at minimizing the risk of disease transmission between animals, as well as between animals and humans. Biosecurity protocols include measures such as quarantine, disinfection, proper waste disposal, control of movement of animals and people, and strict hygiene practices. The goal of biosecurity is to protect the health and welfare of animals, safeguard human health,and prevent economic losses associated with disease outbreaks in livestock.

Biosecurity is often regarded as the first line of defence; if no proper biosecurity is installed on a farm, the efficacy and benefits from other measures such as vaccination or treatments will often be cancelled or reduced. Disease prevention and control typically depend on a holistic set of measures starting from good animal husbandry practices and biosecurity, complemented by specific medical intervention (vaccination, treatment, surgery). In this sense, this book represents a very natural collaboration between two initiatives: BIOSECURE and DISCONTOOLS.

BIOSECURE (biosecure.eu) is a Horizon Europe multi-actor project which seeks to enable decision makers to understand, prioritize and implement evidence-based, cost-effective and sustainable biosecurity management systems. This is to be achieved through a thorough review of existing literature and information surrounding biosecurity across livestock production, quantification of the impact of biosecurity measures and modelling of the risk of disease introduction to the farm, carrying out field studies to assess the efficacy of biosecurity measures, and analysis of the economic impact of biosecurity measures (BIOSECURE, 2023). DISCONTOOLS (discontools.eu), on the other hand, is an EU stakeholder-led initiative to map and prioritize research gaps for infectious diseases of animals. The heart of DISCONTOOLS is an open-access database where expert groups composed of specialists from academia, industry and government bodies have summarized the current state of knowledge and knowledge gaps for 57 infectious diseases of animals. The focus is on the development of required new diagnostics, vaccines, therapeutics or new knowledge to improve infectious disease control in the future.

This book builds on the disease-specific information provided by the DISCONTOOLS expert groups on biosecurity practices and augments it with information collected through an extensive literature search (>500 articles). As such, vets, students, researchers and farm managers will find here a quick overview of the current knowledge on risk factors and biosecurity practices for control and prevention of specific diseases in major livestock species. The early stage researcher and epidemiologist will also find an overview of available epidemiological information and methods to model disease transmission.

Prof. Dr Jeroen Dewulf
Co-ordinator Biosecure consortium

Dr Johannes Charlier
DISCONTOOLS project manager

Introduction

This publication seeks to gather together fundamental information on the transmission and epidemiological characteristics of a range of diseases relevant to the European livestock production. For the purposes of this book, livestock refers to cattle, swine, poultry and small ruminants. We have not considered other productive animal species falling under the umbrella, for example, of aquaculture and game animals. The overall goal of this book is to serve as a brief introductory overview to the availability of key data on these diseases, with a view to informing future work on disease transmission modelling within the context of biosecurity.

We selected the most relevant disease for the European livestock production sector based on a two-step process. Initially, we made use of the DISCONTOOLS (discontools.eu) database to produce an extensive list of potential diseases. We then screened this list for the most relevant diseases for the European context based on EFSA reports, available reviews and prevalence information and/or risk of introduction to Europe. We have not, however, considered the zoonotic importance of the diseases but have merely considered the direct impact on the livestock sector.

This book is subdivided into three chapters. The first chapter gives a brief introduction to general principles and terminology used in biosecurity and modelling transmission of livestock disease. The remaining two chapters provide a review of available information on selected diseases wherein we gathered information on prevalence, risk factors (including risk of introduction to the farm or risk of introduction to Europe), transmission routes, control measures and biosecurity, and the current availability of modelling data. Chapter 2 covers diseases endemic to Europe and is divided into sections for cattle, swine, poultry, small ruminants and diseases affecting multiple species. Where a disease affects multiple species, information has been provided separately for each species affected. Chapter 3 covers the epidemic diseases. Chapter 3 is not subdivided by species, as many of the diseases covered here affect multiple species, so this chapter is only subdivided by disease. As mentioned above, we consider both diseases for which there are ongoing outbreaks across Europe, such as African swine fever and avian influenza, and those diseases present in neighbouring countries and at risk of introduction to Europe, such as Rift Valley fever.

1 Principles and Terminology in Biosecurity

Abstract

In this chapter, biosecurity is defined and the general principles behind it are explored. Within this context, the various transmission routes by which diseases can spread, host pathogen interactions and their importance for biosecurity considerations are then explored. The chapter further explores the different considerations necessary when dealing with epidemic and endemic livestock diseases, and discusses a range of transmission routes and their implications for biosecurity measures. Finally, the chapter examines disease transmission modelling, covering the different types of model available to the researcher, with a particular focus on their role in understanding livestock disease transmission and control, and how modelling can be employed within the context of biosecurity.

Biosecurity: Definition and Its Importance for the Farm and Society

Biosecurity can, in general terms, refer to practices and measures taken to control the entry and spread of disease onto or within a population (Renault *et al.*, 2022). The World Organisation for Animal Health (WOAH) defines biosecurity as 'The application of a set of management, behavioural and physical measures designed to reduce the risk of introduction, establishment and spread of pathogenic agents to, within and from an animal population'. In the context of livestock farming, this refers to the introduction of disease onto, and the spread within, farms and amongst animal populations (Shortall *et al.*, 2016).

Animal disease can have devastating effects on animal welfare (Tomley and Shirley, 2009), in addition to enormous economic impact as demonstrated by the 2001 foot and mouth disease (FMD) outbreak in the UK and the Netherlands (Thompson *et al.*, 2002; Bouma *et al.*, 2003), and the ongoing African swine fever (ASF) outbreak in Europe (Cwynar *et al.*, 2019). Furthermore, it is now well known that over 60% of emerging infections in humans are of zoonotic origin (Robertson, 2020). Taken together, this highlights the importance of effective implementation of biosecurity measures in the livestock farming sector (Chomyn *et al.*, 2023).

In the most general terms, the aim of biosecurity is to prevent the transmission of infectious disease (Alarcón *et al.*, 2021). Under this general aim, biosecurity can be broadly divided into external and internal biosecurity. *External* biosecurity measures are aimed at preventing the introduction of disease into the farm, herd or flock, whereas *internal* biosecurity focuses on measures aimed at preventing spread between animals within the farm, herd or flock (Dewulf and Van Immerseel, 2018). However, as productive systems vary significantly across countries, and between sectors, these general principles can encompass a wide range of measures targeting specific transmission routes.

Properly applied, biosecurity can also serve as an important avenue to reduce the need for curative drug treatment, and this is particularly relevant in the fight against antimicrobial resistance (Dewulf and Van Immerseel, 2018). Positive associations have been reported between farms with a good level of biosecurity and reduced antimicrobial use (Laanen *et al.*, 2013; Postma *et al.*, 2016; Postma and Dewulf, 2018; Dhaka *et al.*, 2023). Nonetheless, despite the long recognized importance of biosecurity measures, their application has often been inconsistent at best. There are a wide range of factors that can influence uptake, including economic considerations (such as cost of making changes to farm buildings and/or perception of a substantially increased time commitment/workload and/or the perceived feasibility of the proposed measures), the overall size of the farm and/or its type of production

DOI: 10.1079/ 9781800629837.0001

system (such as the level of intensity, outdoor or indoor grazing, etc.), a lack of farmer and/or veterinarian knowledge on the subject, and the impracticality of some of the suggested measures (Chomyn *et al.*, 2023). Thus, there often may be a disconnect between the ideal measures that are proposed by researchers and animal health professionals and what is practical and achievable at the individual farm level. Furthermore, this will vary significantly based on the particular farm and the animal sector.

Biosecurity, however, should not be seen only as a simple set of infection control measures, but rather as a holistic approach encompassing the whole food production chain. Biosecurity also has significant overlap with wildlife management and conservation, as many of the diseases of most concern also affect, and spread via wildlife, such as avian influenza (Anthony *et al.*, 2012) and ASF (Sauter-Louis *et al.*, 2021), to name a few.

Beyond the farm, there is a complex set of societal interactions that influence the efficacy and practicability of biosecurity practices. These can be inextricably linked to political affairs and national security on one hand and current political priorities and perceived public interest on the other. A poignant example of the interplay of these issues can be seen in the ongoing Syrian civil war and the resultant regional refugee crisis. This has negatively affected and disrupted the eradication programme for peste des petits ruminants within the country (Benfield *et al.*, 2023). Furthermore, the large-scale displacement of refugees into surrounding countries has led to a concurrent displacement of livestock. Often, these animals are not vaccinated, which has caused an increase in sheep and goat pox cases (Tuppurainen *et al.*, 2017). The proximity of this conflict to the Balkans and the importance of the small ruminant sector in this region further highlight the significant knock-on effects such issues can create for neighbour countries and the EU as a whole. Within the context of Europe, the 2001 FMD outbreak in the UK led to the imposition of an import ban of British meat into the rest of the EU, at which time the UK was still a member (Shanafelt and Perrings, 2018). Thus, national political factors can have significant effects on biosecurity at the national-political and regional-political levels.

At the national and regional levels, biosecurity-related legislation and policies are co-ordinated by various governmental animal health authorities and thus are highly susceptible to changes in funding and political priorities. On a larger scale, the Food and Agriculture Organization (FAO) and the WOAH are responsible for many biosecurity-related tasks at the international level. Co-ordination of efforts at the international level is an essential cornerstone of effective disease control and biosecurity, and reflects global rather than just national challenges (Njeumi *et al.*, 2012).

One of the most important outcomes to date of such co-ordination is the eradication of rinderpest (Njeumi *et al.*, 2012; Roeder *et al.*, 2013). Rinderpest was one of the deadliest diseases of cattle, originally having spread from Central Asia to Europe due to human migrations and associated livestock movement and trade (Roeder, 2011; Roeder *et al.*, 2013). Initial efforts at eradication in the 1960s led to the imposition of strict movement controls and mass vaccination across a number of countries, with some significant successes. However, despite these efforts, the disease reappeared in Africa in the 1980s, following which the global rinderpest eradication programme was formed (Biosecurity beyond borders, 2021). The implementation of a range of traditional measures such as movement controls and mass vaccination, as well as more innovative measures including the use of improved diagnostics, phylogenetic analysis to take account of virus evolution and the development and deployment of new, more heat-stable vaccines finally culminated in its eradication in 2011 (Roeder *et al.*, 2013).

A further important consideration for modern biosecurity is the issue of transboundary diseases. Transboundary diseases are defined as highly transmissible epidemic diseases which have the capacity to spread rapidly and globally and the potential to cause enormous economic and social impact and negatively impact public health (Beltran-Alcrudo *et al.*, 2019; Clemmons *et al.*, 2021). They are of particular concern to all sectors of livestock production (Clemmons *et al.*, 2021). As globalization increases connectivity between countries and across continents and industries, this further raises the risk of disease introduction and rapid disease spread (Beltran-Alcrudo *et al.*, 2019). Transboundary diseases are a particular threat to global food security directly, through the loss of animal products due to mortality (including due to culling for control measures) and loss of production due to morbidity, and due to knock-on effects such as disruption to trade and supply chains (Clemmons *et al.*, 2021). Different sectors are vulnerable to a range of different and potentially devastating transboundary diseases, such as bluetongue, FMD, lumpy skin disease, Rift Valley fever in ruminants (Clemmons *et al.*, 2021), African and classical swine fever (CSF) for swine (Beltran-Alcrudo *et al.*, 2019; Clemmons *et al.*, 2021) and avian influenza in poultry (Clemmons *et al.*, 2021). There are also a number of diseases of particular concern which have recently been introduced in border regions like lumpy skin disease and peste des petits ruminants (Tuppurainen *et al.*, 2017; Whittle *et al.*, 2023).

Transmission Routes

The transmission route of a specific disease can have important consequences for the implementation of biosecurity measures, particularly in the case of multiple transmission routes. Thus, this section will briefly introduce the main concepts of disease transmission, which are considered in more detail in Chapters 2 and 3 with regard to the specific diseases covered herein.

Although biosecurity has traditionally been seen as an issue mainly affecting intensive indoor production systems, particularly poultry and pig production, and often focusing on viral and bacterial pathogens (Dunowska, 2018; Dewulf and Van Immerseel, 2018), increasing attention has been paid paid to biosecurity in outdoor grazing systems such as for dairy cattle (Chomyn *et al.*, 2023). These considerations are also of particular relevance to parasitic infections, as well as bacterial and viral infections more commonly considered in biosecurity research to date. However, it is also worth noting that there are a range of unique considerations when addressing microparasite infections vs those of bacterial and viral infections, especially the often more complex life cycle of macroparasites and the presence of free-living stages, making particularly pasture contamination a significant issue.

In general terms, transmission routes can be grouped under the umbrella of horizontal transmission, referring to transmission between any members of a population that are not parent and offspring, or vertical transmission, which refers specifically to transmission from the parent to its offspring, during gestation, parturition or infancy via nursing. Horizontal transmission routes can be defined as air-borne, via droplets, or via direct or indirect contact (Dunowska, 2018). Although it was originally thought that all respiratory pathogens were spread via the air-borne route (Dunowska, 2018), there has been significant recent debate, in large part due to SARS CoV-19, about the precise definition of air-borne and droplet transmission (Rezaei and Netz, 2021). However, the principal difference between the two is the size of the particles, with air-borne transmission typically relying on smaller particles (<5 mM) which aerosolize in the air, whereas droplet transmission typically relies on larger particles (>5 mM) (Dunowska, 2018). Air-borne transmission also typically relies on inhalation into the lung, whereas droplet transmission relies on settling of infectious droplets on the mucous membranes of the animal (Dunowska, 2018). This aspect of disease transmission can have important consequences for biosecurity measures related to air filtration, and the distance over which diseases can be transmitted (centimetres, metres or even kilometres) (Wenke *et al.*, 2017).

Contact transmission occurs due to either close contact (direct transmission) between infected and susceptible animals or through environmental transmission (indirect contact) where the infectious agent contaminates the environment (Dunowska, 2018). In the case of indirect transmission, this typically requires a pathogen that can survive well outside the host and persist in the environment (Dunowska, 2018), and has important implications for biosecurity measures such as cleaning and disinfection (Chowdhury *et al.*, 2020), or a disease vector in the case of vector-borne transmission (Dunowska, 2018). There are a range of important vector-borne diseases, and diseases with potential vector-borne transmission, for the European context, transmitted by a range of different arthropod vectors (Dórea *et al.*, 2016). There are ongoing outbreaks for a number of these diseases such as bluetongue virus, transmitted by *Culicoides* midges (Baylis *et al.*, 2010), and lumpy skin disease transmitted by a wide range of potential vectors such as stable flies, mosquitoes and certain species of ticks (Whittle *et al.*, 2023).

The faecal–oral route is particularly significant for transmission of gastrointestinal pathogens. Animals with diarrhoea have the capacity to contaminate a wide area, with high pathogen loads that can persist for long periods in the environment (Dunowska, 2018).

Within the context of environmental transmission, it is also of value to consider the role played by objects within the farm environment which can serve as a source of infection. These are typically termed fomites and can include clothing and protective equipment such as boots, vehicles and other equipment typically used on farms (Dunowska, 2018). Although veterinary equipment (such as needles for injection) can also be considered under this umbrella, this route may also be termed iatrogenic and constitutes a somewhat separate route of transmission. Fomite transmission has been implicated in a number of important diseases (Rossi *et al.*, 2017) such as FMD (Auty *et al.*, 2019), CSF (Ribbens *et al.*, 2012) and influenza A (Allerson *et al.*, 2013; Hernández-Jover *et al.*, 2015). Indirect transmission can also occur through contaminated water, most often drinking water, although this can also occur through swimming in contaminated water or use of contaminated water for rinsing equipment (Dunowska, 2018). A range of pathogens can be transmitted through contaminated water including parasites such as *Toxoplasma* spp., bacteria such as *Campylobacter* and *E. coli* and viruses such as avian influenza (Dunowska, 2018).

Sexual transmission occurs in the case of disease where the infectious agent is shed in reproductive secretions (Dunowska, 2018). Although this can occur during natural mating, in terms of biosecurity the more concerning route is via artificial insemination (Lockhart *et al.*, 1996; De Ruigh *et al.*, 2006; Dunowska, 2018). Trade in semen is widespread and has been a component of breeding programmes for some time for both cattle (De Ruigh *et al.*, 2006) and pigs (Knox, 2016). However, this can raise issues for biosecurity, particularly if clinical notes are not sufficient, as a number of devastating cattle diseases can be transmitted via artificial insemination such as bovine viral diarrhoea (BVD) (De Ruigh *et al.*, 2006), ASF in pigs (Friedrichs *et al.*, 2022) and *Mycoplasma agalactiae* in small ruminants (Prats-van der Ham *et al.*, 2018). A further danger with pathogens transmissible via artificial insemination is the large distances that the disease can travel through this route (Dunowska, 2018).

Vertical transmission involves the transmission of the pathogen from the mother to the offspring during gestation, birth or hatching (Dunowska, 2018). A number of important pathogens can also be transmitted via this route such as paratuberculosis (Aly and Thurmond, 2005) and BVD (Newcomer *et al.*, 2017) in cattle, and porcine reproductive and respiratory syndrome (PRRS) virus (Pileri and Mateu, 2016), and porcine circovirus-II (Baekbo *et al.*, 2012) in pigs.

The Host, Pathogen and Complex Interactions

Beyond the transmission routes employed by the pathogen, a complex interplay of additional factors can affect disease transmission within a population. Some pathogens rapidly spread and have the capacity to cause widespread mortality, one example being ASF (Blome *et al.*, 2013), or highly pathogenic avian influenza (Lee *et al.*, 2021). Conversely, pathogens such as *Mycobacterium avium* spp. *paratuberculosis* are characterized by a long delay between infection and shedding, and a large number of animals being subclinically infected (Van Metre *et al.*, 2008; Fecteau, 2018). Some pathogens elicit an effective immune response and thus require a constant supply of new, naïve hosts, whereas influenza viruses, for example, mutate and reassort rapidly and display pronounced antigenic drift, facilitating reinfection of the same individuals (Cattoli *et al.*, 2011; Dunowska, 2018; Xu *et al.*, 2022).

Taken together, these considerations all have important implications for the effectiveness of control strategies. This is particularly important when considering the containment of an epidemic and the adoption of widespread culling, as was practised in the 2001 FMD outbreak in the UK (Tildesley *et al.*, 2009). It is therefore equally important to understand the potential for disease spread prior to taking such drastic measures, and further considerations such as animal welfare and economic concerns should also be taken into account.

In addition to the characteristics of the pathogen, a wide range of host factors can also affect disease spread and may have significant implications for control measures. For example, some diseases can affect multiple different species, with highly variable transmission. In the case of FMD, pigs, cattle and sheep are all susceptible to infection although cattle and sheep are more susceptible to aerosol infection than pigs (Auty *et al.*, 2019). However, despite the different transmission routes, pigs have been implicated in the spread of FMD to ruminants via air-borne transmission, due to their capacity to generate aerosols and shed large amounts of the virus (Brown *et al.*, 2022). Other pathogens can cause severe disease in some species but relatively mild or subclinical infections in others (Dunowska, 2018). This can be a particular concern in the case of zoonotic pathogens, such as *Campylobacter* which causes potentially fatal zoonotic gastrointestinal infection in humans (Quilliam *et al.*, 2013) but is relatively ubiquitous in poultry where it does not cause severe symptoms (Perez-Arnedo and Gonzalez-Fandos, 2019). Equally, African swine fever is asymptomatic in warthogs but causes devastating haemorrhagic fever in domestic pigs (Blome *et al.*, 2013).

Host density can also play an important role in disease spread. In the case of African swine fever, wild boar constitute an important reservoir for the disease and contribute to its spread to domestic pigs (Chenais *et al.*, 2019; de la Torre *et al.*, 2022), with farms with low levels of external biosecurity being most at risk (de la Torre *et al.*, 2022). Consequently, an increase in the density of wild boar populations has been implicated in an increased risk for African swine fever transmission to domestic pigs (Podgórski *et al.*, 2020). In the case of bovine respiratory disease complex, particularly in intensive farming conditions such as those practised in North America, density of cattle, for example in a feedlot, can have a significant impact on disease transmission (Murray *et al.*, 2016). These considerations, along with the potential for wildlife movements, are important when designing control strategies for livestock diseases.

Finally, animal transport practices can lead to an increased risk of contracting certain diseases, particularly due to stresses on the animals (Baptista *et al.*, 2017; Cirone *et al.*, 2019). Preventing animal stress due to overcrowding, social isolation and social instability within herds can also play an important role in mitigating disease spread and improving animal welfare in general (Proudfoot and Habing, 2015).

Transmission Modelling and Disease Control

At its core, a disease transmission model is fundamentally a representation and simplification of a complex real-world system of disease transmission which aims to improve understanding of disease dynamics. It can help to inform public health decision making, particularly for control strategies such as vaccination, movement restrictions or other intervention measures (Kirkeby *et al.*, 2020).

Modelling of disease transmission has a long history. In 1766, the Swiss mathematician Daniel Bernoulli developed a model for the effect of smallpox eradication on mortality within the population, with many recognizable features of modern models such as a clearly defined force of infection parameter and case fatality rate (Dietz and Heesterbeek, 2000). However, it was some time after this before models became established within human medicine, with some of the first modern models dating from the early 20th century (Lessler and Cummings, 2016). These models, named Ross–McDonald models after their originators, focused on malaria transmission, have had significant implications for control strategies that have been implemented over the years. These early models were instrumental in demonstrating that total elimination of the vector was not necessary to block transmission, but that an increase in mosquito mortality itself would be sufficient to drive eradication (Reiner *et al.*, 2015). These models have also been instrumental in the justification of the use of insecticides as an effective form of malaria control (Smith *et al.*, 2012).

Transmission models have not just been important in developing understanding of the transmission dynamics of single diseases, although this has been a significant contribution, but have also been instrumental in the development of concepts such as herd immunity and establishing vaccination thresholds necessary for effective control of epidemics and eradication of diseases (Lessler and Cummings,

2016). A wide range of diseases have been modelled, including viral, bacterial, unicellular and multicellular parasitic diseases in both humans and animals (Rose *et al.*, 2015; Kirkeby *et al.*, 2020). However, modelling a disease requires availability of information for that disease, be it epidemiological information such as prevalence over time and/or ideally information on its transmission dynamics such as the probability of an infection occurring given a direct/indirect contact between an infectious host and a susceptible host. The more information is available, the more accurate a model can be and the better it can be validated. This naturally also depends on the quality of the information and its applicability to the particular model in question. For example, transmission may be affected by factors such as host population density, vector population density or climatic conditions. Furthermore, in some cases, where information is not available, certain parameters may need to be assumed, assigning a value that represents a 'best guess' based on available information.

Due to the inherent complexity of biological systems and the wide range of diseases with markedly different pathologies and transmission dynamics, there are also a wide range of available modelling tools and methodologies. In addition, disease transmission models can range from the early susceptible-infectious (SI) type models, which cover the transition between different disease states modelled by differential equations with a relatively small number of required parameters, to the significantly more complex agent-based models where each epidemiological unit is modelled individually (Mancy *et al.*, 2017). In practice, this often refers to an individual host but an individual unit in agent-based models can also mean, for example, an individual farm. The choice of individual unit is wholly dependent on the question being addressed by the model.

If models consider time, then the model can be termed dynamic, which can be further subdivided into discrete or continuous processes. Continuous time models, as the name suggests, model the change in disease states of the individual units in an approximation of continuous time, although continuous disease states are also possible to model. In discrete time models, fixed predetermined time intervals are chosen, which the model progresses through, and changes to the number of units within each disease state are updated at each time interval until the end of the period simulated (Kirkeby *et al.*, 2020).

Models can further incorporate stochasticity. Typically, random processes are modelled in continuous or discrete time models where state transitions occur at random times based on constant parameters. In discrete time models, this often involves the use of a per time step probability, which is determined by constant parameters. In a probabilistic model, distributions can be used rather than constant values for input parameters, which leads to a variation in output each time the model is run (Kirkeby *et al.*, 2020).

In each case, a number of parameters are needed to calculate the transition between states/compartments within the model. In order for the model to have predictive value, it is further necessary to base model parameters on available data (Carpio and Pierret, 2022). Creating a model with sufficient representation of a real-life system to have predictive value thus requires significant knowledge regarding the system (Kirkeby *et al.*, 2020). Prior to beginning the process of building the model, it is therefore necessary to gather as much information as possible, and if key data are missing then the best course of action needs to be determined, whether that be further experimental work (Kirkeby *et al.*, 2020) or the possibility of inferring parameters from existing data (Hu *et al.*, 2017).

References

Alarcón, L.V., Allepuz, A. and Mateu, E. (2021) Biosecurity in pig farms: A review. *Porcine Health Management* 7(1), 5.

Allerson, M.W., Cardona, C.J. and Torremorell, M. (2013) Indirect transmission of influenza A virus between pig populations under two different biosecurity settings. *PloS One* 8(6), e67293.

Aly, S.S. and Thurmond, M.C. (2005) Evaluation of *Mycobacterium avium* subsp *paratuberculosis* infection of dairy cows attributable to infection status of the dam. *Journal of the American Veterinary Medical Association* 227(3), 450–454.

Anthony, S.J., St Leger, J.A., Pugliares, K., Ip, H.S., Chan, J.M. *et al.* (2012) Emergence of fatal avian influenza in New England harbor seals. *Bacterial Quorum Sensing and Microbial Community Interactions* 3(4), 12.

Auty, H., Mellor, D., Gunn, G. and Boden, L.A. (2019) The risk of foot and mouth disease transmission posed by public access to the countryside during an outbreak. *Frontiers in Veterinary Science* 6, 381.

Baekbo, P., Kristensen, C.S. and Larsen, L.E. (2012) Porcine circovirus diseases: A review of PMWS. *Transboundary and Emerging Diseases* 59, 60–67.

Baptista, A.L., Rezende, A.L., Fonseca, P. de A., Massi, R.P., Nogueira, G.M. *et al.* (2017) Bovine respiratory disease complex associated mortality and morbidity rates in feedlot cattle from southeastern Brazil. *Journal of Infection in Developing Countries* 11(10), 791–799.

Baylis, M., Parkin, H., Kreppel, K., Carpenter, S., Mellor, P.S. *et al.* (2010) Evaluation of housing as a means to protect cattle from *Culicoides* biting midges, the vectors of bluetongue virus. *Medical and Veterinary Entomology* 24(1), 38–45.

Beltran-Alcrudo, D., Falco, J.R., Raizman, E. and Dietze, K. (2019) Transboundary spread of pig diseases: The role of international trade and travel. *BMC Veterinary Research* 15(1), 64.

Benfield, C.T.O., Legnardi, M., Mayen, F., Almajali, A., Cinardi, G. *et al.* (2023) Peste des petits ruminants in the Middle East: Epidemiological situation and status of control and eradication activities after the first phase of the PPR global eradication program. *Animals* 13(7), 1196.

Biosecurity beyond borders (Editorial) (2021) *Nature Food* 2(7), 449–449.

Blome, S., Gabriel, C. and Beer, M. (2013) Pathogenesis of African swine fever in domestic pigs and European wild boar. *Virus Research* 173(1), 122–130.

Bouma, A., Elbers, A.R.W., Dekker, A., Koeijer, A., Bartels, C. *et al.* (2003) The foot-and-mouth disease epidemic in The Netherlands in 2001. *Preventive Veterinary Medicine* 57(3), 155–166.

Brown, E., Nelson, N., Gubbins, S. and Colenutt, C. (2022) Airborne transmission of foot-and-mouth disease virus: A review of past and present perspectives. *Viruses* 14(5), 1009.

Carpio, A. and Pierret, E. (2022) Parameter identification in epidemiological models. In: Agarwal, P., Nieto, J. and Torres, D. (eds) *Mathematical Analysis of Infectious Diseases*. Elsevier, Amsterdam, pp. 103–124.

Cattoli, G., Milani, A., Temperton, N., Zecchin, B., Buratin, A. *et al.* (2011) Antigenic drift in H5N1 avian influenza virus in poultry is driven by mutations in major antigenic sites of the hemagglutinin molecule analogous to those for human influenza virus. *Journal of Virology* 85(17), 8718–8724.

Chenais, E., Depner, K., Guberti, V., Dietze, K., Viltrop, A. *et al.* (2019) Epidemiological considerations on African swine fever in Europe 2014–2018. *Porcine Health Management* 5(1), 10.

Chomyn, O., Wapenaar, W., Richens, I.F., Reyneke, R.A., Shortall, O. *et al.* (2023) Assessment of a joint farmer–veterinarian discussion about biosecurity using novel social interaction analyses. *Preventive Veterinary Medicine* 212, 105831.

Chowdhury, S., Azziz-Baumgartner, E., Kile, J.C., Hoque, M.A., Rahman, M.Z. *et al.* (2020) Association of biosecurity and hygiene practices with environmental contamination with influenza A viruses in live bird markets, Bangladesh. *Emerging Infectious Diseases* 26(9), 2087–2096.

Cirone, F., Padalino, B., Tullio, D., Capozza, P., Lo Surdo, M. *et al.* (2019) Prevalence of pathogens related to bovine respiratory disease before and after transportation in beef steers: Preliminary results. *Animals* 9(12), 1093.

Clemmons, E.A., Alfson, K.J. and Dutton, J.W. (2021) Transboundary animal diseases, an overview of 17 diseases with potential for global spread and serious consequences. *Animals* 11(7), 2039.

Cwynar, P., Stojkov, J. and Wlazlak, K. (2019) African swine fever status in Europe. *Viruses* 11(4), 310.

de la Torre, A., Bosch, J., Sánchez-Vizcaíno, J.M., Ito, S., Muñoz, C. *et al.* (2022) African swine fever survey in a European context. *Pathogens* 11(2), 137.

De Ruigh, L., Bosch, J.C., Brus, M.C., Landman, B. and Merton, J.S. (2006) Ways to improve the biosecurity of bovine semen. *Reproduction in Domestic Animals* 41(4), 268–274.

Dewulf, J. and Van Immerseel, F. (eds) (2018) General principles of biosecurity in animal production and veterinary medicine. In: *Biosecurity in Animal Production and Veterinary Medicine: From Principles to Practice*. CABI, Wallingford, pp. 63–76.

Dhaka, P., Chantziaras, I., Vijay, D., Bedi, J.S., Makovska, I. *et al.* (2023) Can improved farm biosecurity reduce the need for antimicrobials in food animals? A scoping review. *Antibiotics* 12(5), 893.

Dietz, K. and Heesterbeek, J.A. (2000) Bernoulli was ahead of modern epidemiology. *Nature* 408(6812), 513–514.

Dórea, F.C., Elbers, A.R.W., Hendrikx, P., Enoe, C., Kirkeby, C. *et al.* (2016) Vector-borne disease surveillance in livestock populations: A critical review of literature recommendations and implemented surveillance (BTV-8) in five European countries. *Preventive Veterinary Medicine* 125, 1–9.

Dunowska, M. (2018) Circles of disease transmission. In: Dewulf, J. and Van Immerseel, F. (eds) *Biosecurity in Animal Production and Veterinary Medicine: From Principles to Practice*. CABI, Wallingford, pp. 27–62.

Fecteau, M.-E. (2018) Paratuberculosis in cattle. *Veterinary Clinics of North America* 34(1), 209–222.

Friedrichs, V., Reicks, D., Hasenfuß, T., Gerstenkorn, E., Zimmerman, J.J. *et al.* (2022) Artificial insemination as an alternative transmission route for African swine fever virus. *Pathogens* 11(12), 1539.

Hernández-Jover, M., Schemann, K., East, I.J. and Toribio, J.-A. (2015) Evaluating the risk of avian influenza introduction and spread among poultry exhibition flocks in Australia. *Preventive Veterinary Medicine* 118(1), 128–141.

Hu, B., Gonzales, J.L. and Gubbins, S. (2017) Bayesian inference of epidemiological parameters from transmission experiments. *Scientific Reports* 7(1), 16774.

Kirkeby, C., Brookes, V.J., Ward, M.P., Dürr, S. and Halasa, T. (2020) A practical introduction to mechanistic modeling of disease transmission in veterinary science. *Frontiers in Veterinary Science* 7, 546651.

Knox, R.V. (2016) Artificial insemination in pigs today. *Theriogenology* 85(1), 83–93.

Laanen, M., Persoons, D., Ribbens, S., Jong, E., Callens, B. *et al.* (2013) Relationship between biosecurity and production/antimicrobial treatment characteristics in pig herds. *Veterinary Journal* 198(2), 508–512.

Lee, D.-H., Criado, M.F. and Swayne, D.E. (2021) Pathobiological origins and evolutionary history of highly pathogenic avian influenza viruses. *Cold Spring Harbor Perspectives in Medicine* 11(2), a038679–a038679.

Lessler, J. and Cummings, D.A.T. (2016) Mechanistic models of infectious disease and their impact on public health. *American Journal of Epidemiology* 183(5), 415–422.

Lockhart, A.B., Thrall, P.H. and Antonovics, J. (1996) Sexually transmitted diseases in animals: Ecological and evolutionary implications. *Biological Reviews of the Cambridge Philosophical Society* 71(3), 415–471.

Mancy, R., Brock, P.M. and Kao, R.R. (2017) An integrated framework for process-driven model construction in disease ecology and animal health. *Frontiers in Veterinary Science* 4, 155.

Murray, G.M., Cassidy, J.P., Clegg, T.A., Tratalos, J.A., McClure, J. *et al.* (2016) A retrospective epidemiological analysis of risk factors for a primary necropsy diagnosis of bovine respiratory disease. *Preventive Veterinary Medicine* 132, 49–56.

Newcomer, B.W., Chamorro, M.F. and Walz, P.H. (2017) Vaccination of cattle against bovine viral diarrhea virus. *Veterinary Microbiology* 206, 78–83.

Njeumi, F., Taylor, W., Diallo, A., Miyagishima, K., Pastoret, P.P. *et al.* (2012) The long journey: A brief review of the eradication of rinderpest. *Revue Scientifique et Technique (International Office of Epizootics)* 31(3), 729–746.

Perez-Arnedo, I. and Gonzalez-Fandos, E. (2019) Prevalence of *Campylobacter* spp. in poultry in three Spanish farms, a slaughterhouse and a further processing plant. *Foods* 8(3), 111.

Pileri, E. and Mateu, E. (2016) Review on the transmission porcine reproductive and respiratory syndrome virus between pigs and farms and impact on vaccination. *Veterinary Research* 47(1), 108.

Podgórski, T., Borowik, T., Łyjak, M. and Woźniakowski, G. (2020) Spatial epidemiology of African swine fever: Host, landscape and anthropogenic drivers of disease occurrence in wild boar. *Preventive Veterinary Medicine* 177, 104691.

Postma, M., Backhans, A., Collineau, L., Loesken, S., Sjölund, M. *et al.* (2016) Evaluation of the relationship between the biosecurity status, production parameters, herd characteristics and antimicrobial usage in farrow-to-finish pig production in four EU countries. *Porcine Health Management* 2, 9.

Postma, M. and Dewulf, J. (2018) Biosecurity and its relationship with health, production and antimicrobial use. In: Dewulf, J. and Van Immerseel, F. (eds) *Biosecurity in Animal Production and Veterinary Medicine: From Principles to Practice*. CABI, Wallingford, pp. 77–94.

Prats-van der Ham, M., Tatay-Dualde, J., Ambroset, C., De la Fe, C. and Tardy, F. (2018) The moderate drift towards less tetracycline-susceptible isolates of contagious agalactia causative agents might result from different molecular mechanisms. *Veterinary Microbiology* 220, 39–46.

Proudfoot, K. and Habing, G. (2015) Social stress as a cause of diseases in farm animals: Current knowledge and future directions. *Veterinary Journal* 206(1), 15–21.
Quilliam, R.S., Cross, P., Prysor-Williams, A., Edwards-Jones, G., Salmon, R.L. *et al.* (2013) Subclinical infection and asymptomatic carriage of gastrointestinal zoonoses: Occupational exposure, environmental pathways, and the anonymous spread of disease. *Epidemiology and Infection* 141(10), 2022–2021.
Reiner, R.C.Jr., Le Menach, A., Kunene, S., Ntshalintshali, N., Hsiang, M.S. *et al.* (2015) Mapping residual transmission for malaria elimination. *ELife* 4, e09520.
Renault, V., Humblet, M.-F. and Saegerman, C. (2022) Biosecurity concept: Origins, evolution and perspectives. *Animals* 12(1), 63.
Rezaei, M. and Netz, R.R. (2021) Airborne virus transmission via respiratory droplets: Effects of droplet evaporation and sedimentation. *Current Opinion in Colloid & Interface Science* 55, 101471.
Ribbens, S., Goris, N., Neyts, J. and Dewulf, J. (2012) Classical swine fever outbreak containment using antiviral supplementation: A potential alternative to emergency vaccination and stamping-out. *Preventive Veterinary Medicine* 106(1), 34–41.
Robertson, I.D. (2020) Disease control, prevention and on-farm biosecurity: The role of veterinary epidemiology. *Engineering* 6(1), 20–25.
Roeder, P.L. (2011) Rinderpest: The end of cattle plague. *Preventive Veterinary Medicine* 102(2), 98–106.
Roeder, P., Mariner, J. and Kock, R. (2013) Rinderpest: The veterinary perspective on eradication. *Philosophical Transactions of the Royal Society of London. Series B, Biological Sciences* 368(1623), 20120139.
Rose, H., Wang, T., van Dijk, J. and Morgan, E. (2015) GLOWORM-FL: A simulation model of the effects of climate and climate change on the free-living stages of gastro-intestinal nematode parasites of ruminants. *Ecological Modelling* 297, 232–245.
Rossi, G., Smith, R.L., Pongolini, S. and Bolzoni, L. (2017) Modelling farm-to-farm disease transmission through personnel movements: From visits to contacts, and back. *Scientific Reports* 7(1), 2375.
Sauter-Louis, C., Forth, J.H., Probst, C., Staubach, C., Hlinak, A. *et al.* (2021) Joining the club: First detection of African swine fever in wild boar in Germany. *Transboundary and Emerging Diseases* 68(4), 1744–1752.
Shanafelt, D.W. and Perrings, C. (2018) The effect of the post 2001 reforms on FMD risks of the international live animal trade. *EcoHealth* 15(2), 327–337.
Shortall, O., Ruston, A., Green, M., Brennan, M., Wapenaar, W. *et al.* (2016) Broken biosecurity? Veterinarians' framing of biosecurity on dairy farms in England. *Preventive Veterinary Medicine* 132, 20–31.
Smith, T., Ross, A., Maire, N., Chitnis, N., Studer, A. *et al.* (2012) Ensemble modeling of the likely public health impact of a pre-erythrocytic malaria vaccine. *PLoS Medicine* 9(1), e1001157.
Thompson, D.K., Muriel, P., Russell, D., Osborne, P., Bromley, A. *et al.* (2002) Economic costs of the foot and mouth disease outbreak in the United Kingdom in 2001. *Revue Scientifique et Technique de l'OIE* 21(3), 675–687.
Tildesley, M.J., Bessell, P.R., Keeling, M.J. and Woolhouse, M.E.J. (2009) The role of pre-emptive culling in the control of foot-and-mouth disease. *Proceedings. Biological Sciences* 276(1671), 3239–3248.
Tomley, F.M. and Shirley, M.W. (2009) Livestock infectious diseases and zoonoses. *Philosophical Transactions of the Royal Society of London. Series B, Biological Sciences* 364(1530), 2637–2642.
Tuppurainen, E.S.M., Venter, E.H., Shisler, J.L., Gari, G., Mekonnen, G.A. *et al.* (2017) Review: Capripoxvirus diseases: Current status and opportunities for control. *Transboundary and Emerging Diseases* 64(3), 729–745.
Van Metre, D.C., Tennant, B.C. and Whitlock, R.H. (2008) Infectious diseases of the gastrointestinal tract. In: Divers, T. and Peek, S. (eds) *Rebhun's Diseases of Dairy Cattle*. Elsevier, Amsterdam, pp. 200–294.
Wenke, C., Pospiech, J., Reutter, T., Truyen, U. and Speck, S. (2017) Efficiency of different air filter types for pig facilities at laboratory scale. *PLoS One* 12(10), e0186558.
Whittle, L., Chapman, R. and Williamson, A.-L. (2023) Lumpy skin disease – an emerging cattle disease in Europe and Asia. *Vaccines* 11(3), 578.
Xu, C., Zhang, N., Yang, Y., Liang, W., Zhang, Y. *et al.* (2022) Immune escape adaptive mutations in hemagglutinin are responsible for the antigenic drift of Eurasian avian-like H1N1 swine influenza viruses. *Journal of Virology* 96(16), e0097122.

2 Endemic Diseases of Livestock

Abstract

In this chapter, endemic diseases of livestock are examined. Diseases are subdivided into those affecting poultry, swine, cattle, small ruminants, and multiple species. For each disease, a brief overview covering clinical signs, economic impact and the significance of the disease to its respective sector is provided. An overview of data relating to the prevalence of the disease is then provided, with a focus on Europe where data are available. Following this, an overview of risk factors for disease introduction or disease spread within a herd/flock or farm are provided. Transmission routes for the disease are then explored, followed by an overview of current information regarding control measures and biosecurity. Finally, an overview of modelling data is provided, with a focus on the availability of existing models of transmission and experimental work delineating parameters of interest for modellers.

Section 1: Poultry

Coccidiosis

Overview

Coccidiosis ranks as one of the most important endemic diseases in poultry, causing estimated losses of up to €11.9bn in the sector globally (Blake *et al.*, 2020). Coccidiosis is one of the key factors affecting overall performance of broiler production in Europe (Van Limbergen *et al.*, 2020). The causative agents of coccidiosis are obligate intracellular parasites belonging to the phylum *Apicomplexa*, of which various species of *Eimeria* are the most significant to the poultry industry (Mesa-Pineda *et al.*, 2021), of varying aetiologies depending on the specific species (Mesa-Pineda *et al.*, 2021). Infection can lead to deficiencies in nutrient absorption, decreased growth rates, and increased morbidity and mortality (Chapman, 2014). Control measures exist, with management practices at the farm level, vaccines and chemotherapy being the most effective and frequently used (Shivaramaiah *et al.*, 2014). However, frequent and heavy use of anticoccidials has led to widespread resistance documented against all currently used drugs (Noack *et al.*, 2019).

Prevalence

Coccidiosis is considered to be ubiquitous in poultry production globally (Mesa-Pineda *et al.*, 2021). Within Europe, prevalence of *Eimeria* spp. has been examined in Romania (Györke *et al.*, 2013) and Greece (Andreopoulou *et al.*, 2022, Andreopoulou *et al.*, 2023). A worldwide systematic review of the genotypes of *T. gondii* infecting *Gallus gallus* may also be of relevance here (Rodrigues da Costa *et al.*, 2020).

Risk factors

A range of risk factors have been reported for coccidiosis, primarily the quality of management within the farm, the use of floor systems over cage systems (although the latter is not considered acceptable in Europe any longer due to animal welfare concerns), in addition to overcrowding, and accumulation of faeces (Wondimu *et al.*, 2019). An association has been drawn between the age and sex of the chicken, with young birds and male birds thought to be more at risk, although this is still under debate (Wondimu *et al.*, 2019).

Transmission

Transmission is faecal–oral, with the infection easily being transmitted through the ingestion of sporulated oocysts which contaminate the environment. There are a number of factors which contribute to the spread of the parasite, including poor biosecurity and hygiene of personnel and housing for poultry (Wondimu *et al.*, 2019; Greenacre and Morishita, 2021).

Biosecurity for Livestock Diseases in Europe (Alistair Antonopoulos *et al.*)
DOI: 10.1079/ 9781800629837.0002

Movement of personnel between farms has been identified as one of the key drivers of transmission (Greenacre and Morishita, 2021).

Control measures and biosecurity

There are a variety of effective control measures available for coccidiosis: vaccination, prophylactic treatment with anticoccidials (although this is declining in popularity due to growing resistance), feed additives and improved handling and management practices (Mesa-Pineda *et al.*, 2021).

Given the ubiquity of coccidiosis at the time of writing, strict biosecurity measures leading to coccidiosis-free flocks are considered difficult to attain in practice (Peek and Landman, 2011). However, a number of biosecurity measures are recommended to avoid the introduction of the causative agents. *Eimeria* spp. exhibit high reproductive capacity and can lead to significant environmental contamination. Oocyst walls are durable and prevent the parasite from desiccation and provide protection from chemical disinfection.

Availability of modelling data

There are few papers which examine the modelling of *Eimeria* spp. transmission or spread but studies have considered the effects of various disinfectants on oocyst survival (Owen, 1983; Naciri *et al.*, 2011), in addition to various environmental risk factors and prevalence in the UK (Hermans and Morgan, 2007) and the role of wild birds in maintaining biosecurity (Gerhold *et al.*, 2010). Some studies have examined the infective dose of oocyst of different species of *Eimeria* (Williams, 1973, 2001; Jenkins *et al.*, 2017).

Poultry red mite

Overview

Dermanyssus gallinae, or the poultry red mite (PRM), constitutes a major threat to the poultry industry. *D. gallinae* is the most common ectoparasite in poultry farming (Chauve, 1998; Decru *et al.*, 2020). It is estimated to cause annual losses of €31m in Europe (Van Emous, 2005). It is also the cause of significant welfare concerns for laying hens, leading to stress, aggression, anaemia and occasionally death (Pavlicevic *et al.*, 2019).

Prevalence

Dermanyssus gallinae is ubiquitous in poultry farming in Europe. Infestation rates of hen housing average at 83% across Europe, ranging from a low of 11% in Norway, to 34% in Denmark up to 94% in the Netherlands, Belgium and Germany (George *et al.*, 2015). *D. gallinae* was further detected in 95% of farms in Portugal (Waap *et al.*, 2019).

Risk factors

Optimal temperatures in the range of 10–35°C and a high relative humidity of >70% contribute to the reproduction and development of *D. gallinae* (Nordenfors *et al.*, 1999). The tendency of the mites to seek shelter in substructures within housing facilities makes them particularly difficult to eradicate (George *et al.*, 2015). Smooth materials and environments are unfavourable for *D. gallinae*, so cracks in housing and crevices constitute a risk factor for the pest (Decru *et al.*, 2020).

Transmission

Infestations can become established in breeder flocks, rearing and layer farms. Spread can occur through pullet, egg, and manure transport (Mul *et al.*, 2009; Sparagano *et al.*, 2009). Mites are primarily brought into facilities by the actions of people, but can also be brought by other vertebrate animals such as pets or mice (Decru *et al.*, 2020). *D. gallinae* can also survive on dead poultry. Mul and Koenraadt (2009) provide a list of hazards and associated risk levels for transmission covering environmental, employee/personnel and ventilation. Finally, manure constitutes a potential source for reinfestation and should be stored at a distance from housing (Decru *et al.*, 2020).

Control measures and biosecurity

Biosecurity measures for *D. gallinae* typically revolve around preventing the introduction of the pest into the poultry flock. There is limited information available on biosecurity measures specifically and solely geared towards *D. gallinae*, with most approaches focusing on integrated pest management procedures described in control measures and by Decru *et al.* (2020). However, in general, good biosecurity practices, such as limiting and controlling the entry of personnel, correct design of facilities with separation of manure and carcasses from housing areas (Sparagano *et al.*, 2014), and ensuring that contaminated material does not enter the facility during commercial exchange (Roy and Buronfosse, 2011) can have a significant impact on reducing the chance of infestation.

Decru *et al.* (2020) recommend an exhaustive 16-step cleaning process involving sequential steps removing manure, dry, then wet, cleaning, disinfection, pressure washing and final debris removal. Heat treatments (45°C+being lethal for *D. gallinae*) cannot be used when a flock is present in the housing, but can be used in between (Mul *et al.*, 2009). Acaricides can only be used during empty periods and should not be used preventatively (Decru *et al.*, 2020).

Availability of modelling data

There is limited information which would serve as suitable for modelling, except for detailed information on the survival of *D. gallinae* in response to various conditions such as temperature, moisture and disinfectants within the context of controlling infestation (Decru *et al.*, 2020).

Section 2: Swine

Porcine circovirus type II

Overview

Porcine circovirus type II (PCV2) is considered to be ubiquitous, and is probably present in most pig herds. PCV2 is one of the causative agents of a number of syndromes in pigs, the most significant of which is postweaning multisystemic wasting syndrome (PWMS) (Baekbo *et al.*, 2012). PCV2 is also involved in reproductive disorders (Sanchez *et al.*, 2001), porcine respiratory disease complex (PRDC) (Kim *et al.*, 2003), enteritis (Kim *et al.*, 2004) and porcine dermatitis nephropathy syndrome (PDNS). Generally, the presence of other pathogens such as porcine parvovirus (PPV), porcine respiratory syncytial virus (Allan *et al.*, 1999; Harms *et al.*, 2001) and bacterial pathogens such as *Mycoplasma hyopneumoniae* (Opriessnig *et al.*, 2004) enhances the infection with PCV2 and progression to PWMS. An outbreak of PWMS within a herd leads to increases in mortality among weaners of 4–8%, finishers of 2–3.7% in Denmark and Spain (Lopez-Soria *et al.*, 2005; Nielsen *et al.*, 2008) and as high as 10.3% in an earlier study in Denmark (Baekbo *et al.*, 2012). PWMS herds also show lower daily weight gains compared with unaffected herds (Nielsen *et al.*, 2008).

The following section will focus largely on PCV2 and its manifestation of PWMS.

Prevalence

Postweaning multisystemic wasting syndrome in Europe first occurred in France (LeCann *et al.*, 1997) and in the following decade, Europe experienced an epidemic of PWMS (Vigre *et al.*, 2005). There followed a subsequent decrease in reports of PWMS and laboratory-confirmed PCV2 in most of Western Europe (Baekbo *et al.*, 2012).

Risk factors

Studies indicate that a shift in PCV2 genotype may lead to a new outbreak of PWMS at the regional (Vigre *et al.*, 2005; Gagnon *et al.*, 2007; Dupont *et al.*, 2008) and farm level (Cortey *et al.*, 2011). Herds being affected by porcine reproductive and respiratory syndrome (PRRS), other herds being affected by PWMS in the area, the purchase of large amounts of gilts, a herd size >400 sows, a high seroprevalence of PCV2 antibodies in the herd, PPV antibodies in finishers, active PPV in pregnant dams, on-farm semen collection and visitors to the farm without a 3 day pig-free period prior to the visit have all been identified as risk factors for developing PWMS (Baekbo *et al.*, 2012).

Transmission

Porcine circovirus type II has been detected in both nasal and rectal swabs (Grau-Roma *et al.*, 2009), and is shed in similar quantities from nasal, oral and faecal routes up to 209 days post farrowing (Patterson *et al.*, 2011). PCV2 has also been detected in colostrum (Shen *et al.*, 2010), indicating the possibility of both vertical and horizontal transmission (Baekbo *et al.*, 2012). Some experiments in boars also suggest that PCV2 can be shed in semen (Grasland *et al.*, 2009; Madson *et al.*, 2009). Transmission can also occur during close mingling of healthy pigs from PWMS herds with pigs from PWMS-free herds, particularly when in close proximity to each other, such as sharing pens or in neighbouring pens (Jaros *et al.*, 2006; Kristensen *et al.*, 2009).

Control measures and biosecurity

There are a number of vaccines that, since their introduction, have been shown to successfully limit the effects of PCV2, with significant improvements in daily weight gain and significant reductions in mortality (Kristensen *et al.*, 2011). A high level of external biosecurity is key to preventing a herd becoming affected by PWMS. Particularly effective measures include quarantine for purchased animals, changing of boots and clothes at the farm entrance room, delivery of finishers through a bespoke delivery room, a long empty period for sows and weaners, keeping dry sows in collective pens and the vaccination of sows against atrophic rhinitis (Baekbo *et al.*, 2012), in addition to the source of semen used (Madec *et al.*, 2008). Internal biosecurity measures such as limiting pen sizes, hygiene in weaning and farrowing facilities, and hygiene related to preventing parasite infection at the herd level are also important (Madec *et al.*, 2008).

Availability of modelling data

A time-dependent transmission model for PCV2 transmission rates using data from serial transmission experiments is available (Andraud *et al.*, 2009a), in addition to studies examining the effect of husbandry and control measures on PCV2 dynamics (Andraud *et al.*, 2009b), within herd transmission dynamics (Andraud *et al.*, 2008a, 2010), estimation of the transmission rate parameter (Andraud *et al.*, 2008b) and quantitative experimental examination of PCV2 transmission (Rose *et al.*, 2009). Another study examined the effects on PCV2 transmission of feeding of spray-dried plasma to weanlings (Pujols *et al.*, 2011). The role of on-farm biosecurity in small holdings in Uganda has also been considered (Dione *et al.*, 2018).

Recent studies have further examined the role of vaccination on the presence of PCV2 in the environment (López-Lorenzo *et al.*, 2019) and the efficacy of vaccination with both *M. hyopneumoniae* and PCV2 (Boulbria *et al.*, 2021). A cross-sectional study examined risk factors (Alarcon *et al.*, 2011), with a subsequent study looking at the economic efficiency of different control strategies (Alarcon *et al.*, 2013). Finally, the role of coinfection with parvovirus has also been examined (Lagan Tregaskis *et al.*, 2021).

Porcine reproductive and respiratory syndrome

Overview

Porcine reproductive and respiratory syndrome (PRRS) is a viral disease caused by the PRRS virus (PRRSV) affecting domestic pigs. As the name suggests, it manifests as two distinct aetiologies: reproductive failure and pneumonia, the latter of which can also lead to increased susceptibility to secondary bacterial infections. PRRS has one of the highest global economic impacts in pig production (Lunney *et al.*, 2010). It is estimated that farm-level median losses (depending on the level of impact of PRRS on a farm) for PRRS infections are in the region of €74,181–75,724 (Nathues *et al.*, 2017; Renken *et al.*, 2021), with losses of approximately €255 per sow per year (Renken *et al.*, 2021).

Prevalence

There is a lack of general surveillance programmes, so true prevalence data are often not available, but it has been estimated, based on the EU cost-action EuroPRRSnet (www.cost.eu/actions/FA0902/), that some 20–25% of herds could be positive in Denmark and Romania, 50–75% in Greece, Germany and Austria, and as many as 80–95% in Italy and Spain (Baekbo and Kristensen, 2015). A 2014 survey of veterinary practitioners across Europe reported that PRRS was present in 71% of sows and 68% of weaned/growing pigs, with clinical disease occurring in 17% and 23% respectively. Incidence by country, however, varies significantly, with 47% of infected sows reported to exhibit clinical signs in Italy, 22% in Belgium, 10–20% in France, Germany, Netherlands, Denmark, Spain, Austria, Italy and Poland, and 4% in Russia (De Paz *et al.*, 2015). Seroprevalence for PRRSV has been estimated at 50% from a cohort of pig farms in Ireland (Rodrigues da Costa *et al.*, 2020). In Hungary, a national eradication programme reduced seropositivity of herds from 4.7% in 2013–2014 down to 0.96% by 2018 (Nemes *et al.*, 2019).

Only four countries are considered PRRSV free in Europe: Norway, Sweden, Finland (Baekbo and Kristensen, 2015) and Switzerland (Baekbo and Kristensen, 2015; Rappe *et al.*, 2016).

Risk factors

Increased density has been shown to be correlated with an increase in risk for PRRS outbreaks (Arruda *et al.*, 2017; Sanhueza *et al.*, 2020), in addition to a seasonal increase in risk in the summer (Sanhueza *et al.*, 2020). There are also general considerations for the site of the pig production facility such as altitude, terrain and vegetation which can constitute risk factors (Arruda *et al.*, 2017; Jara *et al.*, 2021).

Transmission

In practical terms, PRRS is often spread by the movement of infected animals, with movement of infected pigs into the production facility, transport by contaminated trucks and use of infected semen being the most common routes (Pileri and Mateu, 2016). However, vertical transmission can occur, with piglets born to infected dams capable of being viral shedders. Furthermore, the virus can be found in urine, faeces and semen, in addition to contaminated feed, which can also be a source of transmission for PRRSV (Pileri and Mateu, 2016; Dee *et al.*, 2018, 2021). Aerosol spread has also been documented (Dee *et al.*, 2009; Otake *et al.*, 2010; Arruda *et al.*, 2019), with a potential range of up to 9.1 km (Dee *et al.*, 2021).

Control measures and biosecurity

Vaccination efforts are complicated by a high degree of viral heterogeneity, which makes fully reliable protection difficult to achieve under field conditions (Zhou *et al.*, 2021; Franzo *et al.*, 2022). Biosecurity practices are therefore one of the most effective means of preventing PRRS outbreaks and infections (Silva *et al.*, 2018, 2019). They furthermore comprise an essential component of vaccination plans, preventing the introduction of new strains to which cross-protective immunity is not conferred (Franzo *et al.*, 2022).

Standard procedures include measures such as showering in and out, visitor control, disinfection and decontamination of supplies coming into the facility, truck wash requirements, veterinarian time between farms and management of mortality (Havas *et al.*, 2022). Measures addressing swine movements, transmission by air and water, and movements of people into/within facilities should be prioritized (Silva *et al.*, 2018). Low biosecurity implementation is correlated with a higher incidence of circulating PRRSV (Papatsiros *et al.*, 2022). There is also a significant body of evidence showing the efficacy of air filtration in preventing transmission of PRRS (Dee *et al.*, 2005, 2006; Pitkin *et al.*, 2009a; Alonso *et al.*, 2013a, b; Havas *et al.*, 2022).

As PRRSV can contaminate feed, there is a growing body of research examining the role of feed mitigants, such as formaldehyde-based solutions and various acids, in reducing PRRSV infections (Dee *et al.*, 2021; Havas *et al.*, 2022).

Availability of modelling data

Stochastic modelling of within-herd transmission has been carried out (Evans *et al.*, 2010), in addition to infection dynamics (Clilverd *et al.*, 2023) and time-series models examining temporal trends in Danish swine herds (Antunes and Jensen, 2019). Studies have modelled transmission of PRRS and the effects of vaccination (Galvis *et al.*, 2022a). Additional transmission routes such as animal movements (Makau *et al.*, 2021), vehicle movements (VanderWaal *et al.*, 2020; Galvis *et al.*, 2022b) and feed ingredients has also been modelled (Galvis *et al.*, 2022b). A number of studies have focused on between-farm transmission routes (Galli *et al.*, 2022), air-borne spread (Pitkin *et al.*, 2009b) and dissemination across pig production systems (Jara *et al.*, 2021). A further study examined within-host response to exposure dose, duration and virulence levels (Go *et al.*, 2019) and viraemia profiles under experimental infection (Islam *et al.*, 2013, 2019).

Desrosiers (2011) examined PRRS transmission within the wider context of the transmission of swine pathogens. A Poisson-based model examined recirculation detection (Fraile *et al.*, 2020). Further studies examined the introduction, persistence and fade-out of PRRS within a breeding herd (Nodelijk *et al.*, 2000), effectiveness of vaccination (Nodelijk *et al.*, 2001), population dynamics (Nodelijk *et al.*, 2003), the quantitative risk of introduction into New Zealand via infected pig meat (Cobb *et al.*, 2015) and the validation of a scoring system to assess vulnerability of breeding herds to PRRSV introduction (Silva *et al.*, 2018). Scenario tree modelling of surveillance systems (Frössling *et al.*, 2009) and vaccination (Rathkjen and Dall, 2017; Bitsouni *et al.*, 2019) for eradication have been studied.

Management factors and strategies (David *et al.*, 2021; Campler *et al.*, 2022) and risk factors in general have been examined (Mortensen *et al.*, 2002; Velasova *et al.*, 2012). Fablet *et al.* (2016) further examined herd-level PRRSV seroconversion rates and collected data on herd characteristics, biosecurity measures, management, climatic and housing conditions. Machine learning algorithms have been developed to identify biosecurity practices and risk factors associated with PRRS (Silva *et al.*, 2019). Economic efficiency of control strategies has also been examined (Nathues *et al.*, 2018).

Finally, an expert review of the perceived efficacy of on-farm biosecurity measures related to a number of viral pathogens of swine and cattle also considered PRRS (Kuster *et al.*, 2015).

Swine *mycoplasma*

Overview

Mycoplasma hyopneumoniae is a major pathogen of the porcine lower respiratory tract. Endemically infected herds typically experience high degrees of morbidity but low degrees of mortality due to *M. hyopneumoniae* pneumonia (Pieters and Maes, 2019). This often results inreduced growth rates (Ferraz *et al.*, 2020) and increased feed conversion ratios (Pieters and Maes, 2019). A variable proportion of animals can remain subclinically infected (Garcia-Morante *et al.*, 2022). Naive populations, however, canlead to more severe clinical signs and mortality (Bargen, 2004), with the severity often influenced by the strain of *M. hyopneumoniae* (Woolley *et al.*, 2012; Michiels *et al.*, 2017), coinfection with other respiratory pathogens (Saade *et al.*, 2020), genetics (Borjigin *et al.*, 2016) and environmental conditions (Michiels *et al.*, 2015). Although losses are mainly in the grow-finishing pigs, breeding animals, where infections are typically asymptomatic, can transmit the infection to their offspring in the farrowing unit (Pieters *et al.*, 2009, 2014).

Prevalence

Most herds are endemically infected with *M. hyopneumoniae* but the presence in breeding animals of the bacterium and associated antibodies varies between farms. In a study in Belgium, PCR prevalence ranged from 0 to 43.8% and seroprevalence from 32.5 to 93.8% at the farm level. A significant amount of farm-level seroprevalence variance is attributable to gilt vaccination (Biebaut *et al.*, 2022). An earlier study examining pig herds in Belgium, Netherlands, Portugal, Spain, France, Germany and Italy found that 42.7% of herds tested by PCR were positive for *M. hyopneumoniae*, with 5.6% of individual piglets tested being positive (Labarque and Trinidade, 2012). The genotype of *M. hyopneumoniae* also differs significantly between regions (Dos Santos *et al.*, 2015).

Risk factors

Purchases of >120 gilts per year, circulating respiratory pathogens in the breeding population and *M. hyopneumoniae*-positive sows within the farrowing unit have been identified as risk factors for increased prevalence at weaning (Nathues *et al.*, 2013; Pieters *et al.*, 2014). The *M. hyopneumoniae* infection status of the dam also plays a significant role in the likelihood of transmission (Pieters *et al.*, 2014). Mixing and sorting of pigs is a further risk factor, as it stresses the animals and increases disease transmission likelihood (Maes *et al.*, 2008; Pieters and Maes, 2019). Finally, strain differences can account for differences in disease severity, so certain strains may constitute a greater risk than others (Vicca *et al.*, 2003).

Transmission

Although typically born *M. hyopneumoniae* free, piglets can become infected within the first week of life (Sibila *et al.*, 2007; Holst *et al.*, 2015). Infected sows constitute a source of colonization for piglets (Calsamiglia and Pijoan, 2000). Shedding of the dam during lactation has been shown to significantly influence the probability of *M. hyopneumoniae* in piglets (Pieters *et al.*, 2014). Although indirect transmission by fomites has been proposed (Goodwin, 1985), this is likely to be less important than direct transmission routes (Maes *et al.*, 2008).

Control measures and biosecurity

Strict hygiene measures and restricting the movement of equipment and personnel between groups of animals of different ages are important (Maes *et al.*, 2008). Management practices such as all-in/all-out (AIAO) production, maintaining good herd immunity and limiting destabilizing factors, careful optimization of stocking densities, prevention of other respiratory diseases, optimal housing and optimal climatic conditions have been identified as important for disease control (Maes *et al.*, 2008). AIAO production is probably the most important factor, as this can interrupt the cycle of transmission from older animals to offspring. AIAO further allows for implementation of uniform conditions and for cleaning of facilities between populations (Maes *et al.*, 2008). Quarantine or an adaptation period of >30 days is recommended when purchasing animals (Amass and Baysinger, 2006). A stocking density of 0.7 m^2 per pig in fully slatted floors has been recommended (Madec and Josse, 1984).

Tetracyclines and macrolides are frequently used for *M. hyopneumoniae* control, although there are concerns about growing resistance. Finally, the use of commercial vaccines has been shown to improve daily weight gain, feed conversion and mortality rate (Maes *et al.*, 2008).

Availability of modelling data

Meyns *et al.* (2004) quantified the adjusted reproduction ratio (R_n) of *M. hyopneumoniae* in piglets during their nursery period (4–10 weeks of age). An experimental study examined the role of exposure route on *M. hyopneumoniae* infection in pigs, covering intratracheal, aerosol and intranasal routes, and collected data on a number of body score indices such as weight and cough index; body fluid samples were collected throughout (Silva *et al.*, 2021). Desrosiers (2011) examined *M. hyopneumoniae* transmission within the wider context of the transmission of swine pathogens. Browne *et al.* (2017) examined the *in vitro* survival of *M. hyopneumoniae* in the context of the role of dust and low temperature and pig housing, while environmental contamination by *M. hyopneumoniae* has also been explored (Garza-Moreno *et al.*, 2022). Nathues *et al.* (2016) examined within-herd transmission in closed herds. Kruse *et al.* (2020) conducted multivariate analysis of biosecurity, productivity and antimicrobial and vaccine use in Danish sow herds.

Finally, the role of coinfections in *M. hyopneumoniae* outbreaks has been examined in a field study in Italy (Tonni *et al.*, 2022). Fablet *et al.* (2016) further examined the role of *M. hyopneumoniae* infections in the context of herd-level PRRSV seroconversion rates, and collected data on herd characteristics, biosecurity measures, management, climatic and housing conditions.

Swine pleuropneumonia

Overview

Actinobacillus pleuropneumoniae is the aetiological agent of swine pleuropneumonia, a contagious respiratory disease of domestic pigs causing significant economic losses globally (Gottschalk and Broes, 2019). Estimations of reduction in daily weight gain from the disease range from 0.77 to 25.5% (Holmgren *et al.*, 1999). In its acute manifestation, it leads to haemorrhagic and necrotizing pleuropneuomonia which is often fatal (Sassu *et al.*, 2018).

Prevalence

Swine pleuropneumonia is widely distributed, with major outbreaks occurring in practically every European country (Gottschalk and Broes, 2019).

Risk factors

Environmental stressors or infection with other respiratory pathogens can lead to sudden outbreaks of clinical disease (Gottschalk and Broes, 2019). *A. pleuropneumonae* is divided into multiple serovars which can vary significantly in their virulence, with some exhibiting high mortality whereas others produce few, if any, symptoms. This is often determined by the production of various toxins by the different serovars (Frey, 1995). Within-host factors can also play a significant role in determining the risk of severe disease, such as the presence or absence of protective antibodies and coinfection with other pathogens (Sassu *et al.*, 2018).

Transmission

In endemically infected herds, transmission is typically to offspring from infected sows during close contact. Large quantities of *A. pleuropneumoniae* can be shed in nasal discharges from infected animals (Gottschalk and Broes, 2019). During acute outbreaks, evidence suggests that aerosol transmission can occur over short distances (Jobert *et al.*, 2000; Savoye *et al.*, 2000; Desrosiers *et al.*, 2004), and can occur over longer distances within buildings and via farm personnel (Gottschalk and Broes, 2019). Asymptomatic carriers, either animals surviving an outbreak or those with subclinical infection, can become a source of infection for naïve animals (Klinkenberg *et al.*, 2014). *A. pleuropneumoniae* survival in the environment is typically short in warm and dry conditions but mucus or other organic matter can prolong survival to several days to weeks, with survival in clean water possible for up to 39 days at 4°C (Gottschalk and Broes, 2019). Although common disinfectants are effective, organic matter must first be removed (Gutiérrez *et al.*, 1995).

Control measures and biosecurity

There are a number of commercial vaccines available, classified by type as bacterins (washed and killed whole bacteria); purified toxin based; and mixed bacterins and toxin based for specific serovars. For piglets, first doses should not be applied prior to 7 weeks of age to prevent maternal antibody interference, and age adjustment should be carried out depending on when clinical signs appear, depending on the serovar. Boosters are also highly recommended and vaccination should be avoided when there are concurrent high levels of PRRSV transmission within the herd. When vaccinating sows and/or gilts, two doses should be applied, with the second 3 weeks prior to farrowing. Vaccination can also be carried out on replacement gilts during quarantine. Care should be taken as vaccination may not eliminate bacteria colonizing the tonsils.

Beyond vaccination and antimicrobial treatment, the use of AIAO systems, batch farrowing, control of environmental conditions such as temperature and ventilation, and the control of other infections that can predispose animals to severe disease (such as *M. hyopneumoniae*) are effective in reducing *A. pleuropneumoniae* (Sassu *et al.*, 2018; Gottschalk and Broes, 2019).

Availability of modelling data

Assavacheep and Rycroft (2013) examined survival of *A. pleuropneumoniae* outside the pig, studying persistence in various solutions, wet and dry states,and different temperatures. Fablet *et al.* (2016) examined the role of *A. pleuropneumoniae* within the wider context of herd-level PRRSV seroconversion rates, and collected data on herd characteristics, biosecurity measures, management, climatic and housing conditions. A number of studies have also examined the herd factors and characteristics associated with seroprevalence of different serovars of *A. pleuropneumoniae* (Maes *et al.*, 2001) and seroconversion rates (Vigre *et al.*, 2004). Several studies have examined the role of different air filtration systems for pig facilities (Wenke *et al.*, 2017,

2018). Finally, the dynamics of polymicrobial infections within the wider context of porcine respiratory disease complex following African swine fever introduction has been examined (Assavacheep and Thanawongnuwech, 2022).

Section 3: Cattle

Paratuberculosis

Overview

The causative agent of paratuberculosis, also known as Johne's disease, is *Mycobacterium avium* subsp. *paratuberculosis* (MAP), causing a chronic granulomatous infection of the GI tract in domestic and wild ruminants. It displays a long incubation period, followed by delayed faecal shedding and humoral response in infected animals (Fecteau, 2018). Given the long incubation period, some 95% of cattle are considered to be subclinically infected (Van Metre *et al.*, 2007). Clinical disease generally manifests as diarrhoea and weight loss (Fecteau, 2018). Economic modelling suggests that losses due to MAP and Johne's disease range from €51.5m and €68.5m in France and Germany respectively (Rasmussen *et al.*, 2021).

Prevalence

In an extensive review of paratuberculosis in 48 countries, Whittington *et al.* (2019) found herd-level prevalence <1% to be present only in 2/31 countries which collected prevalence data. A majority of the 31 countries (58%) which collected prevalence data showed >20% of farmed ruminant herds were infected. Herd prevalence was >40% in beef and dairy cattle in Brittany, France, 20–40% and 10–20% respectively in The Netherlands and UK, >40% in beef cattle in Germany and Belgium, 20–40% in dairy cattle in Belgiun and Ireland, >40% in dairy cattle in Italy and 10–20% for both beef and dairy cattle in Greece, representing a significant portion of the herds in these countries. Sweden, conversely, showed <1% of herds infected in both beef and dairy cattle. Within-herd prevalence of >10% was common in most countries, with >15% prevalence observed in more than a tenth.

Prevalence is expected to be underestimated in 60% of countries examined due to the types of tests used, lack of surveillance, lack of farmer/veterinarian knowledge, reluctance to report due to government measures and lack of concern from government.

Risk factors

In herds with no previous history of MAP infection, transmission is most likely to occur through purchase of infected animals which may show no signs of infection and test negative for culture and serology (Fecteau, 2018). Infection of dams represents a significant risk to calves (Whittington and Windsor, 2009). MAP survives pasteurization temperatures so the feeding of colostrum or milk from infected cows, or feeding of pooled waste milk and colostrum, serves as a risk factor (Chiodini and Hermon-Taylor, 1993). Replacement stock exposure to contaminated environments and offering feed refusals from adult cattle to younger heifers can also facilitate transmission (Rossiter and Burhans, 1996). Intensive grazing systems can increase the risk of transmission to replacement stock due to pasture contamination (Fecteau and Whitlock, 2011), although typically high infectious doses are required (Fecteau, 2018). Certain breeds may have a genetic predisposition (Kirkpatrick, 2010).

Transmission

Transmission is principally faecal–oral (Fecteau, 2018), most often in calves sucking manure-contaminated teats and udders (NRC, 2003), although vertical transmission to calves is also possible *in utero* (Aly and Thurmond, 2005), as is transmission via colostrum itself and milk during the later stages of infection (Sweeney *et al.*, 1992; Streeter *et al.*, 1995). Aerosol transmission has also been reported as possible (Eisenberg *et al.*, 2012). Semen from subclinically infected bulls can be a source of transmission (Ayele *et al.*, 2004). Beef calves are more likely than dairy calves to be exposed to adults shedding MAP, whereas for dairy calves physical separation or removal to a heifer-raising facility lowers the risk (Fecteau, 2018). MAP is resistant to acidic soil, low and freezing temperatures, and can survive for up to 55 weeks in a dry shaded environment (Whittington *et al.*, 2004).

Control measures and biosecurity

The most effective control measures are long-term herd management strategies, which can be subdivided into two overall categories: (1) ensuring negative herds remain so by avoiding contact with infected animals either by introduction from other herds or participation at auctions/shows where there may be exposure; (2) reducing the prevalence within infected herds, preventing new infections, managing infected cattle properly and improving resistance of animals to MAP. Test and cull (Lu *et al.*, 2008; McAloon *et al.*, 2019), colostrum management and maintenance of clean and segregated calving and calf-rearing areas to reduce the possibility of MAP faecal contamination (Doré *et al.*, 2012) are important (Fecteau, 2018). Chemoprophylactic administration of antimicrobials such as gallium has been shown to inhibit MAP growth in neonatal calves (Fecteau and Whitlock, 2011). Vaccination is effective in reducing faecal shedding and disease onset (Bastida and Juste, 2011; Groenendaal *et al.*, 2015).

A number of control programmes exist, beginning in France in 1920, Netherlands in 1942, the UK and Norway in the 1960s, followed by Cyprus in the 1970s (Benedictus *et al.*, 2000). However, despite expansion of programmes to Ireland, Spain, Italy, Switzerland, Czechia and Denmark, overall there is a lack of co-ordinated information available for national control programmes at the time of writing (Whittington *et al.*, 2019).

Availability of modelling data

A number of studies have examined the modelling of various relevant aspects of paratuberculosis control and transmission. Wells and Wagner (2000) presents a general model of biosecurity in relation to dairy production in the US. Several studies have examined between-herd (Biemans *et al.*, 2022) and within-herd (Biemans *et al.*, 2021) transmission of MAP, infection dynamics (Smith *et al.*, 2015; Al-Mamun *et al.*, 2016) and the efficacy of national control programmes in Ireland (Jordan *et al.*, 2020) and Italy (Scarpellini *et al.*, 2023). Studies have also examined the spread and surveillance of MAP in the Swedish cattle trade network (Rosendal *et al.*, 2020), and the role of cattle trade movements in control (Ezanno *et al.*, 2022).

Control measures to prevent increases in paratuberculosis (Ben Romdhane *et al.*, 2017; Camanes *et al.*, 2018), including at the regional (France) (Beaunée *et al.*, 2017) and herd level (Chile) (Verdugo *et al.*, 2018), have been considered, in addition to the economic costs of control (Garcia and Shalloo, 2015; Smith *et al.*, 2017; Chiu *et al.*, 2018). Several studies have looked at aspects such as farmer compliance (Kirkeby *et al.*, 2018), voluntary testing (Kanankege *et al.*, 2019), farm characteristics and farmer perceptions (Broughan *et al.*, 2016) and the role of calf rearing technology in paratuberculosis (Vass-Bognár *et al.*, 2022). Environmental aspects such as soil pH (Cowled *et al.*, 2016) and calving areas (Field *et al.*, 2023) have been studied.

Finally, stochastic simulation has been used to examine the use of vaccination to prevent invasion of MAP (Lu *et al.*, 2013).

Bovine mycoplasmosis

Overview

Mycoplasma bovis is the causative agent of bovine mycoplasmosis of global significance to the cattle industry (Gelgie *et al.*, 2022). It is responsible for a number of different aetiologies including mastitis, pneumonia, otitis, arthritis and endocarditis (Hale *et al.*, 1962; Gourlay *et al.*, 1976; Foster *et al.*, 2009; Hananeh *et al.*, 2018; Kanda *et al.*, 2019). Mastitis is one of the most significant pathologies, causing widespread economic loss and animal welfare issues (Gelgie *et al.*, 2022). *M. bovis* pneumonia is also a pathology of significant interest, but is often considered under the umbrella of bovine respiratory disease (BRD) (Bernal *et al.*, 2023). There is growing cause for concern due to declining antibiotic efficacy (Gautier-Bouchardon *et al.*, 2014).

Prevalence

Prevalence varies significantly between regions. A longitudinal study from Ireland based on bulk milk tank sampling reported a herd-level prevalence of *M. bovis* of between 0.42% and 53%, depending on the year (McCarthy *et al.*, 2021). In Sweden, prevalence was estimated at 4.8% (Hurri *et al.*, 2022), whereas in Estonia it was estimated at 48.3% (Mõtus *et al.*, 2021). An earlier study from Portugal estimated prevalence in the north-west of the country to be 2.4% based on culture results (Pinho *et al.*, 2013). A high prevalence of *M. bovis* is also reported in England and Wales, being the most frequently identified species in cattle between 2005 and 2019 (Deeney *et al.*, 2021).

Risk factors

Introduction of diseased animals via the purchasing of cattle (Fox *et al.*, 2005; Schibrowski *et al.*, 2018; Murai and Higuchi, 2019; Fujimoto *et al.*, 2020) and the use of an infected breeding bull or contaminated semen (Gille *et al.*, 2018; Haapala *et al.*, 2018) are significant risk factors for the introduction of *M. bovis* to a herd. Poor hygiene practices are also a significant risk for the spread of disease (Fox *et al.*, 2005).

Transmission

Transmission occurs from cows to calves via milk consumption (Aebi *et al.*, 2012), with shedding of *M. bovis* during parturition (Gille *et al.*, 2018) constituting another potential source. Nose-to-nose contact is another source of transmission, either via diseased animals or asymptomatic carriers (Fox, 2012). Haematogenous spread from one portion of the body to another, and systemically to different organs, is also possible (Biddle *et al.*, 2005; Houlihan *et al.*, 2007). *M. bovis* spreads rapidly to all age groups once it enters a herd (Vähänikkilä *et al.*, 2019), with horizontal transfer between pens reported (Timsit *et al.*, 2012).

Control measures and biosecurity

Identification of asymptomatic carriers of *M. bovis* is difficult (Maunsell *et al.*, 2011), due in part to the long incubation period (Gelgie *et al.*, 2022), so it is necessary to implement effective control measures to prevent introduction to the herd (Fox, 2012; Haapala *et al.*, 2021). Young stock are seen as a potential reservoir for colonization (Aebi *et al.*, 2015), with newborn animals acquiring the infection from older ones within the herd. Calves, 1–4 months of age, generally show the highest prevalence, dropping between 8 and 12 months old, with the key to stopping the chain of infection being preventing spread to calves born following first detection of *M. bovis* (Haapala *et al.*, 2021). This can effectively be accomplished by raising calves separately from older animals (Aebi *et al.*, 2015).

Further control measures include culling and isolating *M. bovis* mastitic cows (Punyapornwithaya *et al.*, 2012), pasteurizing infected milk (Butler *et al.*, 2000) and in extreme cases culling all cattle (Pothmann *et al.*, 2015).

More general principles such as separation of diseased animals, good hygiene and ventilation, and avoidance of overcrowding (Gorden and Plummer, 2010) are also important in controlling and preventing the spread of bovine mycoplasmosis.

Availability of modelling data

The majority of studies have focused on the modelling of various diagnostic parameters and their role in screening for *M. bovis* (Bokma *et al.*, 2022; Cowled *et al.*, 2022). Recent studies also considered transmission trends during an outbreak in New Zealand (Jordan *et al.*, 2021) and dynamics of within-herd prevalence (Timonen *et al.*, 2017, 2020).

***Staphylococcus aureus* mastitis**

Overview

Staphylococcus aureus is also considered under antimicrobial-resistant (AMR) pathogens, and is listed in a recent EFSA report as one of the most significant pathogens for AMR in cattle, alongside *E. coli* (EFSA AHAW, 2021). In this section, the focus will be on mastitis caused by *S. aureus*, which is the leading cause of mastitis in dairy cattle worldwide (Campos *et al.*, 2022).

Clinical mastitis manifests as swelling and/or tenderness in the udder and visible abnormalities in milk, whereas subclinical mastitis does not manifest overtly; nevertheless, both forms are causes of significant economic losses (Gonçalves *et al.*, 2018; Campos *et al.*, 2022;

Paramasivam *et al.*, 2023). Bacterial infections damage the secretory epithelia, altering the total milk yield and milk components (Le Roux *et al.*, 2003). Estimates of losses can be as high as 1.4 kg per quarter per milking (Gonçalves *et al.*, 2018). Bovine *S. aureus* contamination in milk and raw milk products is also a significant cause of human food-borne illnesses (Fagundes *et al.*, 2010).

Prevalence

Staphylococcus aureus is a commensal and an opportunistic pathogen of both livestock and humans (Fluit, 2012). Overall prevalence is highly variable between regions and herds (Schnitt and Tenhagen, 2020). The general focus on prevalence, therefore, centres on the molecular epidemiology of clones (Campos *et al.*, 2022).

Risk factors

Poor milking hygiene (Barkema *et al.*, 2006; Schnitt and Tenhagen, 2020), herd size and density (Schnitt and Tenhagen, 2020), age of cows (Barkema *et al.*, 2006), larger mammary size (Deluyker *et al.*, 2005) and contact with pigs, as pigs are carriers of *S. aureus* and constitute a reservoir, are potential risk factors for the development of *S. aureus* mastitis (Porrero *et al.*, 2012; Abreu *et al.*, 2019; Schnitt and Tenhagen, 2020). Humans can also be a source of *S. aureus* transmission so lack of appropriate hygiene by farm workers also constitutes a source of risk (Schnitt and Tenhagen, 2020), as does visits of people who have visited multiple farms during the day (Middleton *et al.*, 2002).

Transmission

Older studies from the US estimated that most herds have at least one cow infected with *S. aureus*, with within-herd prevalances reaching as high as 94% in some cases (Dinsmore, 2002). *S. aureus* enters the mammary gland via the teat canal (Schnitt and Tenhagen, 2020). Most often, a single strain is predominant within a herd, spreading from cow to cow (Zadoks *et al.*, 2002; Barkema *et al.*, 2006; Keefe, 2012). Milking constitutes the primary risk period for transmission, with the most common routes being milkers' hands, milking equipment and udder cloths (Schnitt and Tenhagen, 2020).

Control measures and biosecurity

Antibiotic treatment, while an option for individual animals, is not favourable due to a combination of both costs and the risk of antibiotic resistance, and thus cannot be considered as a viable long-term strategy for controlling persistent *S. aureus* infection (Fluit, 2012). Antibiotic use is now strictly controlled in some countries (Gomes and Henriques, 2016). Milking hygiene, particularly use of postmilking disinfectants, is key to controlling the spread of *S. aureus* (Barkema *et al.*, 2006).

Availability of modelling data

There is a variety of modelling data available for *S. aureus* mastitis, including a field-based validation of network models for disease spread between dairy herds (Álvarez *et al.*, 2011), mathematical modelling of *S. aureus* control in dairy herds (Zadoks *et al.*, 2002), molecular and mathematical epidemiology (Zadoks, 2002), the role of lactation therapy in treating subclinical mastitis (Barlow *et al.*, 2009), a dysbiosis model of infection, a gamma frailty model of udder quarter infection dynamics (Ampe *et al.*, 2012), strain effect on colonization site (Exel *et al.*, 2023) and stochastic modelling of the effects of *S. aureus* strains on economic and epidemiological outcomes of mastitis interventions (Exel *et al.*, 2022). Recent studies have examined the prevalence and risk factors for subclinical mastitis in Rwanda (Ndahetuye *et al.*, 2020) and Kenya (Mbindyo *et al.*, 2020). Authors have also reported examination of the epidemiological behaviour of mastitis pathogens in bulk milk tank sampling (Fernández *et al.*, 2013) and a review of mastitis spatial analysis (Gil *et al.*, 2022).

Bovine respiratory disease complex

Overview

Bovine respiratory syncytial virus (BRSV) and bovine herpes virus-I (BHV-I) are two of the most common causative agents of bovine respiratory disease complex (Taylor *et al.*, 2010; Bidokhti *et al.*, 2012). As such, prior to considering these diseases individually, a separate section covering certain aspects of BRD is included here as a whole. Some overlap is present between the section on BRD and those for BRSV and BHV-I but this is unavoidable due to the close association between these viruses and BRD as a whole.

Bovine respiratory disease is a significant cause of economic damage, with a mortality rate of 2.1%, leading to a loss of €1007 per mortality, €234 loss for animals receiving >3 treatments for BRD, and a loss of €40–130 for subclinical infections (Blakebrough-Hall *et al.*, 2020). A variety of factors can influence the severity of BRD, including age and immunological status (Makoschey and Berge, 2021), with symptoms ranging from subclinical to fatal (Ellis *et al.*, 1996). In practice, it is often difficult to attribute clinical signs of BRD to a single pathogen but experimental evidence is available for individual viruses such as BRSV (Makoschey and Berge, 2021).

Prevalence

Bovine respiratory disease is a common problem, and one of the leading causes of serious respiratory infection in dairy and beef cattle globally (Kirchhoff *et al.*, 2014). A nationwide epidemiological study in France found that cumulative incidence of BRD at the farm level was 9.8% (2.1% at the animal level) with an overall mortality of 0.1% (Gay and Barnouin, 2009). A study in Norway carried out in 1999 found that outbreaks of BRD occurred in 19% of herds within the study area (Norström *et al.*, 1999).

Risk factors

A number of risk factors are associated with BRD morbidity, such as age (Townsend *et al.*, 1989) and genetics (Snowder *et al.*, 2006, 2007; Snowder, 2009), in addition to management decisions such as mixing of animals from different sources and transport stress (Baptista *et al.*, 2017; Cirone *et al.*, 2019). Feedlot density and the length of time spent on pasture/in housing are also significant factors (Murray *et al.*, 2016). BRD in Europe is thus more likely to occur during the cold months, when cattle are in housing (Gay and Barnouin, 2009).

Transmission

Bovine respiratory disease can be transmitted by five possible pathways: direct and indirect contact, inhalation, ingestion, vertically and venereally (Renault *et al.*, 2018). Experimental studies have shown that air-borne transmission of viral causative agents may also be possible (Mars *et al.*, 1999).

Control measures and biosecurity

As many of the pathogens related to BRD are endemic, it is in practice difficult to fully eliminate. Management practices should focus on minimizing pathogen shedding, exposure and transmission within the population, but various combinations must be applied at the individual farm level for best effect (Callan and Garry, 2002). Renault *et al.* (2018) identified six distinct categories for controlling the spread of BRD: animal movements, direct contact (internal/external) with shedders/carriers, feed and water contamination, indirect contamination via fomites, and human and environmental contamination. They further identified general management (housing, clean and dry bedding, adequate ventilation, cemented/concrete flooring), general hygiene practices, management of sick animals, and calves and calving management as important to control BRD spread.

Availability of modelling data

The effect of management decisions and housing, including the type of bedding used in barns and calf batch size, has been modelled (Assié *et al.*, 2009), in addition to the efficacy of biosecurity measures at the population level in Norway in the context of the Norwegian national control programme (Stokstad *et al.*, 2020). Modelling of risk factors at the animal or herd level and environmental risk factors has also been conducted in Ireland (Murray *et al.*, 2016) and Austria (Klein-Jöbstl *et al.*, 2015), and transport stress in France (Pratelli *et al.*, 2021), in addition to a comparison of the likelihood of infection on organic vs conventional dairy farms (Bidokhti *et al.*, 2009). The effects of antimicrobial usage and pen size have also been modelled (Picault *et al.*, 2022).

Further studies considered the effect on productivity of cattle production parameters and their relation to BRD incidence in France (Delabouglise *et al.*, 2017). A cluster analysis of BRD identified two distinct epidemiological patterns affecting either young calves during winter or older calves (>5 months) at any time throughout the year (Bernal *et al.*, 2023). Finally, several studies have examined the economic cost of BRD (van der Fels-Klerx *et al.*, 2001; Mohd Nor *et al.*, 2012; Wilson *et al.*, 2017).

Bovine respiratory syncytial virus

Overview

Bovine respiratory syncytial virus (BRSV) is a significant cause of acute respiratory disease in dairy and beef calves globally (Sarmiento-Silva *et al.*, 2012). It is one of the most frequent causative agents of bovine respiratory disease complex (Cusack, 2023), and its aetiology is largely considered under this umbrella. Peak clinical signs manifest 4–6 days post infection, with high (40°C) fevers, depression, anorexia and increased respiratory rate, with airway obstruction occurring due to mucus production (Ciszewski *et al.*, 1991; LeBlanc *et al.*, 1991). Severe disease does occur but most animals recover within 10 days unless coinfection is present (Jolly *et al.*, 2004; Ellis *et al.*, 2010).

Prevalence

Bovine respiratory syncytial virus is widely distributed in cattle globally (Valarcher and Taylor, 2007). In northern Italy, BRSV antibodies were detected in at least one cow in all herds examined, with 25% of herds showing 100% seropositivity (Luzzago *et al.*, 2010). In Poland, BRSV was detected in 60% of herds sampled (Urban-Chmiel *et al.*, 2015). In Sweden, a total of 30 BRSV outbreaks were recorded between 2007 and 2011 (Bidokhti *et al.*, 2012). Seropositive herds ranged from 73 to 82% of those examined between 2011 and 2012 (Wolff *et al.*, 2015), with an earlier study showing bulk milk tank seropositivity of 41–89%, depending on the region of the country (Elvander, 1996). BRSV is responsible for 60% of BRD outbreaks within dairy and beef herds in Norway (Stokstad *et al.*, 2020).

Risk factors

Specific risk factors identified for BRSV beyond the general categories considered above for BRD include the number of incoming movements for individual herds, geographical differences in proximity to neighbouring herds (Ohlson *et al.*, 2010; Toftaker *et al.*, 2016) and herd size (Norström *et al.*, 2000a,b; Ohlson *et al.*, 2010; Toftaker *et al.*, 2016), and not providing boots to visitors (Ohlson *et al.*, 2010).

Transmission

Adult animals with subclinical infection are the main source of infection for within-herd reinfection (Baker and Frey, 1985; Baker *et al.*, 1986). However, a genetic diversity study of BRSV in Denmark suggests that reintroduction into a herd constitutes the major vehicle of transmission at herd level (Larsen *et al.*, 2000). Interherd transmission is common (Bidokhti *et al.*, 2012), with evidence of geographical clustering of genetic sequences suggesting the possibility of air-borne and mechanical transmission routes (Valarcher *et al.*, 2000). Experimental evidence also suggests that an air-borne transmission route may be possible for BRSV (Mars *et al.*, 1999).

Control measures and biosecurity

Good general hygiene and management measures are key to preventing the introduction or reintroduction of BRSV into a herd, in line with general guidelines for BRD (Renault *et al.*, 2018). In terms of specific measures, vaccination against BRSV is one of the most effective means of control (Makoschey and Berge, 2021), although this is not used by the Norwegian control programme, which instead relies on classification of herds based on seropositivity and prevention of introduction through improved biosecurity (Stokstad *et al.*, 2020). Colostrum management has been associated with lower seropositivity for BRSV antibodies (Pardon *et al.*, 2015).

Availability of modelling data

A number of studies have modelled the role of risk factors on seropositivity of herds in Norway (Norström *et al.*, 2000b; Toftaker *et al.*, 2016) and Sweden (Ohlson *et al.*, 2010). Biosecurity measures and their effectiveness in the context of the control programme in Norway have recently been considered (Stokstad *et al.*, 2020). Studies have examined the carriage and spread of BRSV by fomites and human nasal mucosa (Oma *et al.*, 2018), infection dynamics between and within herds (Klem *et al.*, 2013) and viral shedding in experimentally infected calves and its relation to transmission (Klem *et al.*, 2019).

Greenhalgh and Griffiths (2009a) constructed a three-stage epidemic model of BRSV and examined dynamic phenomena arising from an extended core group model (Greenhalgh and Griffiths, 2009b), and finally the probability of extinction in a BRSV epidemic model (Griffiths and Greenhalgh, 2011).

Infectious bovine rhinotracheitis

Overview

Bovine alphaherpesvirus 1 (BoHV-I/BHV-I), the causative agent of infectious bovine rhinotracheitis (IBR), infectious pustular vulvovaginitis (IPV) and infectious pustular balanoposthitis (IPB), can also cause abortion, infertility and encephalitis (Nandi *et al.*, 2009). IBR typically manifests as high fever, anorexia, rapid respiration and purulent nasal discharge (Divers, 2007). As is typical of many herpesviruses, BoHV-I is capable of recrudescence in animals stressed by other infectious diseases, transport stress or medications such as corticosteroids. IBR is generally associated with high levels of morbidity, and although IBR itself has low mortality, when fatalities occur they are generally due to secondary bacterial pneumonia or concurrent viral infection with BRSV or bovine viral diarrhoea virus (BVDV). Vaccination or natural infection tends to offer only short-term protection (6–12 months) (Divers, 2007). IBR is a cause of significant economic losses in the cattle industry (Iscaro *et al.*, 2021), with estimates of up to €62 m annually in Ireland (Animal Health Ireland, 2021).

Prevalence

Bovine alphaherpesvirus 1 has been successfully eradicated from Norway (Stokstad *et al.*, 2020), with Finland, Denmark, Sweden, Austria, northern Italy, Czechia and Germany also being certified as IBR free (Iscaro *et al.*, 2021). In Poland, 53% of herds tested were found to be seropositive (Rypuła *et al.*, 2012), in Ireland it is estimated at 28–42% (Graham, 2013), 39.8% in Slovakia, and as low as 3.6% in Slovenia (Hostnik *et al.*, 2021), whereas there are few data for other countries such as Hungary, or Spain (Iscaro *et al.*, 2021).

Risk factors

Specific risk factors identified for BoHV-I beyond the general categories considered above for BRD include contact with genital secretions or semen of infected animals, treatment with corticosteroids and parturition (Winkler *et al.*, 2000). In addition, many of the general risk factors for BRD, such as transport stress and coinfection with bacterial or viral pathogens (Baptista *et al.*, 2017; Cirone *et al.*, 2019), also constitute specific risk factors for reactivation of latent virus (Iscaro *et al.*, 2021).

Transmission

Nasal discharges, cough droplets, fetal fluids and semen, and genital secretions constitute the main sources of BoHV-I. Transmission generally occurs via direct contact with infected animals or indirectly via contaminated materials or personnel (Iscaro *et al.*, 2021). Due to the potential for latent animals to reactivate infection, these should always be considered as a potential source of BoHV-I (Bitsch, 1973). Air-borne transmission between calves under field conditions is also possible (Mars *et al.*, 1999).

Control measures and biosecurity

Since the 1970s, many European countries have initiated national control and eradication programmes specifically targeting BoHV-I (Iscaro *et al.*, 2021). IBR is currently a notifiable disease only if it occurs in a IBR-free territory or country (European Commission, 2020).

The overall form of control strategies is highly dependent on the economic and epidemiological features of the implementing country. In low BoHV-I seroprevalence areas, for example, culling of all seropositive, unvaccinated animals has been found to be effective (Iscaro *et al.*, 2021), in addition to creating BoHV-I free breeding stock (Ackermann and Engels, 2006). This has been successful in eradicating the condition in Scandinavia, northern Italy and Austria. Vaccination followed by removal of seropositive animals is considered effective in areas of moderate-to-high seroprevalence. Belgium and the Netherlands, for example, have adopted a vaccination-based strategy. Control programmes also vary between mandatory, for example in Scandinavia, Ireland, France, Czechia, Slovakia and Germany, whereas in Italy, for example, eradication plans are voluntary but offer monetary incentives.

Finally, some countries only operate partial control programmes focusing on, in the case of Estonia for example, control of semen (Iscaro *et al.*, 2021) or are only implemented at the regional level in the case of Galicia in Spain (Villaamil *et al.*, 2020).

Availability of modelling data

A recent systematic review examined risk factors for the introduction of BoHV-I into herds, with a focus on collecting inputs to be used for modelling eradication scenarios and for biosecurity communication (Waldeck *et al.*, 2021). Santman-Berends *et al.* (2018) have examined the quantification and probability of IBR reintroduction into the Netherlands via cattle imports. An earlier study modelled the epidemiology and spread of IBR in the Netherlands (Noordegraaf *et al.*, 1998) and an additional study modelled the effect of surveillance programmes on BoHV-I spread between certified cattle herds (Graat *et al.*, 2001). Brock *et al.* (2021) produced a large-scale epidemiological model of BoHV-I spread in Irish cattle. O'Grady *et al.* (2008) modelled risk factors in the context of serosurveillance of bulls at a bull performance station in Ireland. A risk assessment of the introduction of both BVDV and BoHV-I through indirect contacts is also relevant both to IBR and BVD (considered below) (Benavides *et al.*, 2020, 2021).

Other studies have used scenario tree modelling for targeted herd sampling to determine freedom from disease for IBR and enzootic bovine leucosis (Blickenstorfer *et al.*, 2011), and stochastic modelling for freedom from disease for IBR in Switzerland (Audigé *et al.*, 2001). Finally, an expert review of the perceived efficacy of on-farm biosecurity measures related to a number of viral pathogens of swine and cattle also considered IBR (Kuster *et al.*, 2015).

Bovine viral diarrhoea

Overview

Bovine viral diarrhoea virus (BVDV) is the causative agent of bovine viral diarrhoea (BVD), one of the most important and economically impactful diseases of cattle worldwide (Moennig and Becher, 2018). Economic impact can vary by country, with losses typically ranging from lows of €4–25 in Hungary, to €32–69 in Ireland, to upwards of €96 in Germany and France (Richter *et al.*, 2017). BVDV infections can result in a variety of clinical manifestations in cattle of all ages, including the eponymous enteric disease in addition to respiratory disease (wherein it is often considered under the umbrella of BRD; see above) (Fulton *et al.*, 2009). Reproductive and fetal disease can also occur in the case of BVDV infections in breeding females.

Prevalence

A global systematic review and meta-analysis of BVDV prevalence estimates the seroprevalence within Europe as a whole at 45.56%. A pronounced seasonal variation is also seen at the global level, with the highest seroprevalence observed in winter (63.44%) and antigen prevalence observed in spring (21.33%), and the lowest seroprevalence observed in autumn (42.25%) and antigen prevalence in the summer (4.17%). Antigen prevalence in Europe as a whole was estimated at 23.27% (Su *et al.*, 2022).

Risk factors

Risk factors for BVD include large herd sizes, close proximity to neighbouring herds, over-fence contacts, a high number of infected neighbours, heifers on common pasture and purchasing animals without BVD documentation (Lindberg *et al.*, 2006).

Transmission

Bovine viral diarrhoea virus transmission can occur through a variety of routes. Direct transmission typically occurs when the virus is ingested or inhaled (Grooms *et al.*, 2014), and via contact with saliva, oculonasal discharge, uterine secretions, amniotic fluid, placenta, semen, urine and faeces from persistently infected animals (Newcomer *et al.*, 2017). BVDV can cross the placental barrier (Grooms *et al.*, 2014), leading to calves being born tolerant of endogenous virus which then become persistently infected and shed virus without clinical signs of disease (Moennig and Becher, 2018). Indirect BVDV transmission can also occur via vectors such as haematophagous insects or via veterinary equipment (Grooms *et al.*, 2014). BVDV can also persist in the environment for up to 14 days, but is susceptible to common disinfectants. However, the virus can survive for up to 21 days at 5°C in slurry (Bøtner and Belsham, 2012). Transport of persistently infected animals has long been recognized as one of the principal sources of herd-to-herd transmission (Houe *et al.*, 2006).

Control measures and biosecurity

Vaccination efficacy has been hampered by the fact that it does not necessarily prevent vertical transmission to offspring, and thus does not prevent emergence of persistently infected cattle. Thus, vaccination alone cannot eliminate the reservoir for future BVDV infections, which requires constant booster vaccination and further measures (Moennig and Becher, 2018).

Several European countries have instituted national eradication plans: Denmark in 1994 (Bitsch *et al.*, 2000); Switzerland in 2008 (Schweizer *et al.*, 2021); Northern Ireland in 2013 (Strain *et al.*, 2021); and Germany in 2011 (Wernike *et al.*, 2017). National control and eradication programmes in Europe typically rely on the identification and removal of persistently infected animals (Larghi, 2018). Scandinavian countries, Austria and Switzerland have implemented control programmes without vaccination, whereas in Germany, Belgium, Ireland and Scotland, vaccination forms an important optional component of control programmes.

In populations with endemic BVDV, seroprevalence is typically >90% and new infections typically do not lead to severe disease. This presents an issue with control programmes during the final phases. Systematic removal of persistently infected cattle leads to a dramatic reduction in seroprevalence, which in turn presents a major cause for concern for reintroduction. This vulnerability highlights the essential role of strict biosecurity measures (Moennig and Becher, 2018). BVD-specific biosecurity typically centres around preventing herd-to-herd spread, with traceability and identification procedures being of particular importance (Larghi, 2018). Limiting contact with neighbouring herds via over-fence contacts is also a key aspect of BVD control (Lindberg *et al.*, 2006).

Availability of modelling data

Damman *et al.* (2015) modelled the spread of BVDV, examining transmission rates in different housing, risk of external introduction via contact with neighbouring herds and purchase of animals, and the risk of pregnant cattle generating persistently infected calves. A further study examined neighbourhood contacts and trade movements and their role in BVDV spread (Qi *et al.*, 2019). Sarrazin *et al.* (2014) compared the transmission of BVDV by persistently and transiently infected animals, calculating R_0 for the BDVD-1b and BVDV-2a strains respectively, in addition to R_0 for persistently infected animals.

Gates *et al.* (2013) examined the associations of cattle movements, local spread and biosecurity. Chamorro *et al.* (2011) examined vector transmission of BVDV by horn fly. Ersbøll *et al.* (2010) examined spatial modelling of between-herd infection dynamics. Courcoul and Ezanno (2010) modelled the spread of BVDV in a managed metapopulation, and in an earlier study also carried out sensitivity analysis to identify key parameters to model BVDV spread (Ezanno *et al.*, 2007). Viet *et al.* (2007) reviewed assumptions and modelling options for within-herd spread and developed an approach for qualitative validation for stochastic simulation of BVDV within-herd spread (Viet *et al.*, 2006).

Fountain *et al.* (2022a) examined the economic outcomes of a number of on-farm biosecurity measures related to BVD. A further study examined the economics of control programmes (Han *et al.*, 2020). Pinior *et al.* (2019) describe a meta-analysis of epidemiological factors and mitigation measures influencing production loss. A further study reported the preferences of farmers to guide on-farm biosecurity measures (Fountain *et al.*, 2022b). A number of studies have examined aspects of the effectiveness of control programmes (Sekiguchi *et al.*, 2018; Han *et al.*, 2018; Guelbenzu-Gonzalo *et al.*, 2021; Casey-Bryars *et al.*, 2022). A study by van Roon *et al.* (2022) applied a Bayesian hidden Markov model to examine herd-level freedom from infection. Quantitative risk assessment for the introduction of BVDV has also been reported (Foddai *et al.*, 2014; Benavides *et al.*, 2021). Rajeev *et al.* (2017) considered pathogen exposure at the livestock–wildlife interface. A Belgian study examined the benefits of biosecurity measures on farm productivity (Renault *et al.*, 2018). Finally, an expert review of the perceived efficacy of on-farm biosecurity measures related to a number of viral pathogens of swine and cattle also considered BVD (Kuster *et al.*, 2015).

Section 4: Small Ruminants

Small ruminant lentiviruses

Overview

Small ruminant lentiviruses (SRLVs) are highly genetically variable retroviruses infecting sheep and goats. They induce systemic infection which manifests as syndromes which can affect a variety of target organs, including the lungs, nervous system, mammary glands and joints. This largely depends on the strain of SRLV, the species infected and the genetic predisposition of the animal or breed to certain aetiologies. However, only syndromes affecting the nervous or respiratory system lead to death of the animal, although other syndromes can lead to production losses and animal welfare issues (Minguijón *et al.*, 2015).

Prevalence

de Miguel *et al.* (2021) carried out a systematic review and meta-analysis assessing the worldwide prevalence of SRLVs. They reported a mean individual prevalence of 40.9% of sheep tested (n=4,07,509) in Europe since 1981. Greece and Spain show the highest individual prevalence in Europe at 66% and 50% respectively. Flock prevalence varies considerably, from a low of 3% in the UK to 17% in Belgium and 34% in Poland (although a large-scale study later showed herd prevalence may be as high as 61%; Kaba *et al.*, 2023), to 49 and 51% in France and Germany respectively, up to 89% in Spain and 99% in Italy.

Risk factors

Risk factors include the use of winter housing (Minguijón *et al.*, 2015), length of time spent in housing, ventilation of housing, incorporation of new animals into flocks, and high density in pens (Blacklaws *et al.*, 2004; Reina *et al.*, 2009). Intensive production systems have been shown to be a significant risk factor for SRLVs infection and spread (Barquero *et al.*, 2013; Konishi *et al.*, 2016; de Miguel *et al.*, 2021).

Transmission

Although SRLVs can be detected in milk and mammary glands (Bolea *et al.*, 2006), only around 10–15% of infections are accounted for by this transmission route (Herrmann-Hoesing *et al.*, 2010). Horizontal transmission, with animals in direct contact with each other (Leginagoikoa *et al.*, 2006; Pérez *et al.*, 2010) via winter housing facilities, has been posited as the most likely route of transmission based on evidence contrasting extensive outdoor production systems in Texas, Australia and southern Spain vs seasonal housing facilities as used in Iceland, which underwent a relatively recent epidemic outbreak (Minguijón *et al.*, 2015). The high density of ovines in Greece and Spain, which rely on intensive management systems, further supports this (de Miguel *et al.*, 2021).

Although air-borne transmission is possible, it requires close contact and continuous exposure (Villoria *et al.*, 2013). Sexual transmission is considered possible (Kaba *et al.*, 2023) but unlikely to constitute a significant cause of spread (Minguijón *et al.*, 2015).

Control measures and biosecurity

Minimizing susceptible animal exposure to shedders is key to controlling SRLV spread, with periodic test and cull recommended (Pérez *et al.*, 2010). Where this is not feasible, it is recommended that animals be segregated or separated to limit the possibility of horizontal transmission (Leitner *et al.*, 2010). To prevent the possibility of sexual transmission, ideally, only certified SRLV-free males should be used as semen donors. To prevent indirect transmission, it is also recommended that milking devices be maintained and cleaned regularly, in addition to placing known SRLV-free animals at the head of the milking chain. Some control programmes also recommend not sharing equipment to minimize transmission, and quarantine periods prior to introduction of SRLV-free animals to pastures grazed by infected ones (Minguijón *et al.*, 2015).

Concern also exists regarding the potential for wildlife to spread the infection due to the cross-susceptibility of wild ruminant species (Minardi da Cruz *et al.*, 2013). Experimental evidence further suggests a role for faecal contamination of water, with SRLVs also detected in troughs and inside pens (Villoria *et al.*, 2013), and there is a low but possible chance of iatrogenic transmission via needles (Minguijón *et al.*, 2015).

Availability of modelling data

No relevant literature was encountered which undertook modelling of SRLVs.

Contagious agalactia

Overview

Contagious agalactia (CA) is caused by various *Mycoplasma* spp. in small ruminants (Nicholas *et al.*, 2008). Although there is little recent information regarding economic costs (Jaÿ and Tardy, 2019), a historical study estimated the cost to Mediterranean rim countries as €27.4 million per year at the time of writing (Nicholas, 2002). At the herd level, more recent estimates place potential losses at €7,000–130,000 depending on herd size (Loria *et al.*, 2018). CA affects lactating females and young animals (Corrales *et al.*, 2007), where mortality rates can reach 100% (Bergonier *et al.*, 1997), although this is typically less severe in older animals, depending on the species of *Mycoplasma* (Bergonier *et al.*, 1997; Kumar *et al.*, 2014). Infection generally results in rapid spread of acute disorders within the herd over several weeks (Kinde *et al.*, 1994), with mastitis, pyrexia, anorexia and polyarthritis being common clinical signs (Jaÿ and Tardy, 2019).

Prevalence

Contagious agalactia is most often reported in Mediterranean rim countries, and within Europe is prevalent in Spain, with *M. agalactiae* detected in 9–55% of herds, depending on the test used (Assunção *et al.*, 2004; De la Fe *et al.*, 2007; Amores *et al.*, 2012). Prevalence in France was estimated at 2–20% in goat and sheep herds between 2003 and 2008 (Chazel *et al.*, 2010). Reports also show the presence of CA aetiological agents in Greece and Cyprus (Giadinis *et al.*, 2012; Filioussis *et al.*, 2014) and Italy (Fusco *et al.*, 2007; Agnello *et al.*, 2012; Cillara *et al.*, 2015).

Risk factors

Risk factors for the development of CA include the use of outsider rams and generally having contact with other flocks/animals/people, improper cleaning of milking equipment (Al-Momani *et al.*, 2008), in addition to the size of the herd/flock (Matos *et al.*, 2019).

Transmission

Mycoplasma spp. can persist in infected organs and areas such as the ear canal for long periods of time (Jaÿ and Tardy, 2019). Within-herd transmission is principally oral via feeding, and respiratory (Filioussis *et al.*, 2014). Although the environment does not constitute a sustainable reservoir, it may play an indirect role in transmission through shared areas such as troughs, feeders (Bergonier *et al.*, 1997), bedding (Justice-Allen *et al.*, 2010) and milking practices, which are thought to constitute a significant source of transmission (Kinde *et al.*, 1994; Todaro *et al.*, 2015).

Control measures and biosecurity

Contagious agalactia is listed as a notifiable animal disease by the WOAH. Control measures typically revolve around culling of affected flocks, in addition to chemotherapy (Jaÿ and Tardy, 2019), with early, collective and repeated administration of antimicrobials being recommended to eliminate infection (Garrido *et al.*, 1987; Lambert, 1987). Biosecurity measures can generally be considered under the umbrella of mastitis-causing pathogens and their associated biosecurity measures (Barkema *et al.*, 2009). These include having standard operating procedures for biosecurity (Dinsmore, 2002) and preventing the introduction of infected animals (Gunn *et al.*, 2008), particularly in regard to diagnostic testing prior to introduction into the herd/flock.

Ideally, dairy herds/flocks should be closed, with barriers to outside animals and people, although in practice this is not always possible (Barkema *et al.*, 2009). Isolation of affected animals is recommended where possible, although this is often not possible in practice due to rapid spread of infection. Reduction of mammary transmission can be achieved by thorough checking and cleaning of milking equipment (Al-Momani *et al.*, 2008).

Availability of modelling data

Several studies have examined the efficacy of decontamination (Eterpi *et al.*, 2011) and the survival of *Mycoplasma* spp. in the environment (Justice-Allen *et al.*, 2010), and carried out experimental infections to determine key factors for host colonization (Baranowski *et al.*, 2014). However, beyond this, few studies have examined any aspects of contagious agalactia transmission, dynamics or other categories of interest for modelling parameters.

Section 5: Multiple species

Nematodes

Overview

Gastrointestinal nematodes (GINs) of ruminants such as *Haemonchus* spp., *Teladorsagia circumcincta*, *Trichostrongylus colubriformis* in small ruminants, *Ostertagi ostertagia* and *Cooperia oncophora* in cattle, and *Nematodirus* spp. affecting both cattle and small ruminants are a significant cause of production losses and animal welfare issues within the ruminant production sector, causing upwards of €1.8bn in economic losses (Charlier *et al.*, 2020). The lungworm *Dictyocaulus viviparus* is also a significant cause of morbidity in grazing cattle of temperate regions (Schunn *et al.*, 2013). This is further compounded by widespread resistance to most drug classes, including increasing cases of cross- and multidrug resistance (Rose Vineer *et al.*, 2020a). In domestic pigs, *Ascaris suum* and *Trichuris suis* are of particular importance in fatteners and *Oesophagostomum dentatum* in adult pigs (Thamsborg *et al.*, 2013).

Prevalence

Gastrointestinal nematodes are ubiquitous in outdoor systems, but prevalence is generally seasonal and heavily influenced by climatic conditions such as temperature and humidity, which affects egg survival and hatching in the environment (Troell *et al.*, 2005; Ola-Fadunsin *et al.*, 2019; Băieş *et al.*, 2022), although specific conditions vary by species and productive animal type.

All grazing livestock are exposed to GINs, although different species typically are predominant in different biomes, and there is a significant seasonal component to species-specific GIN infections (Besier *et al.*, 2016). The prevalence of GINs of poultry can be as high as 100% in backyard and free-range systems (Rabbi *et al.*, 2006; Sherwin *et al.*, 2013), and the prevalence of GINs of sheep can also be as high as 100% in some cases (Martínez-Valladares *et al.*, 2013), but this is significantly affected by, and can vary according to, seasonal conditions (Eysker *et al.*, 2005; Kenyon *et al.*, 2009). The prevalence of pig nematodes varies considerably from as low as 2–3% up to 100% for certain species in different farms (Băieş *et al.*, 2022; Eslahi *et al.*, 2022), driven primarily by the production system rather than climatic conditions (Eslahi *et al.*, 2022).

Risk factors

The principal risk factor for infection with GINs is common across all productive animal species considered herein, namely outdoor production systems which facilitate the transmission of infective larvae/eggs via faecal–oral contamination of the environment, in addition to specific climatic conditions favouring egg hatching and/or survival (Troell *et al.*, 2005; Martínez-Valladares *et al.*, 2013; Wongrak *et al.*, 2014, 2015; Besier *et al.*, 2016; Ola-Fadunsin *et al.*, 2019; Băieş *et al.*, 2022; Eslahi *et al.*, 2022).

Transmission

Most poultry (Wongrak *et al.*, 2014, 2015), ruminant (Zajac, 2006) and pig (Christensen *et al.*, 1995; Pittman *et al.*, 2010; Holland, 2013) nematodes exhibit a direct life cycle. In general terms, eggs shed in faeces hatch and grow to infective stage larvae that are then ingested, where they mature, sexually reproduce, shed eggs (or in rare cases, such as *D. viviparus*, L_1 larvae) in faeces, which contaminate the environment, and repeat the cycle.

Control measures and biosecurity

Infection with GINs typically varies throughout the year, with seasonal patterns emerging in most production systems. This traditionally allowed standardized practices to be communicated to farmers based on a fixed calendar (Hawkins, 1993), although climate change and changes in land use have complicated this in recent years (Charlier *et al.*, 2018).

Since their inception in the 1960s, anthelmintics from three major drug classes (benzimidazoles, cholinergic agonists and macrocyclic lactones) have been the mainstay of GIN control in livestock (Coles *et al.*, 2006; Charlier *et al.*, 2018; Rose Vineer *et al.*, 2020a), poultry (Wongrak *et al.*, 2015) and swine (Pettersson *et al.*, 2021). Traditionally, mass administration to the whole flock/herd was used widely, but in recent years this has become contraindicated in ruminant production due to the increasing prevalence of resistance (Rose Vineer *et al.*, 2020a). More sustainable approaches include informed administration of anthelmintics based on clinical signs to inform and optimize GIN control. Methods such as targeted treatment, relying on metaphylactic treatment based on diagnostic indicators (Kenyon and Jackson, 2012), and targeted selective treatment, focusing on assessing clinical indicators such as weight gain (Busin *et al.*, 2014), are now favoured (Busin *et al.*, 2014; Charlier *et al.*, 2014; Kupcinskas *et al.*, 2017).

Vaccination is an ongoing area of interest, although at present the only licensed vaccine for GINs of ruminants is Barbervax® against *H. contortus* (Claerebout and Geldhof, 2020). Finally, there is significant interest in the selection of animals which are either resistant (immune system capable of reducing/preventing infection) or tolerant (capable of tolerating worm burdens) to GIN infection (Burke and Miller, 2020).

Availability of modelling data

Rashid *et al.* (2019) carried out a systematic review of modelling approaches to determine economic losses caused by cattle parasites. Several studies have examined population dynamics and the effects of climate and climate change on gastrointestinal helminths on pasture (Rose *et al.*, 2016; Rinaldi *et al.*, 2015; Rose Vineer *et al.*, 2016, 2020b; Wang *et al.*, 2018, 2022), the dissemination of parasites in small ruminant farms (Vasileiou *et al.*, 2015), the effect of weaning age on animal performance in naturally acquired helminth infections (Högberg *et al.*, 2023), novel control strategies and diagnostics (Lurier *et al.*, 2018), targeted and targeted selective treatments (Höglund *et al.*, 2013; Charlier *et al.*, 2014), parasite aggregation (Grenfell *et al.*, 1995), population modelling (Smith and Grenfell, 1994) and transmission (Grenfell *et al.*, 1987; Smith and Grenfell, 1994), including dynamically across the whole life cycle (Filipe *et al.*, 2023). Studies have also examined the association between milk production and infection with *D. viviparus* (lungworm) in the Netherlands (Dank *et al.*, 2015; May *et al.*, 2018) and diagnosis (Lurier *et al.*, 2018).

However, there is significantly less information suitable for modelling on poultry helminths, with most studies focusing on prevalence and risk factors for poultry in backyard flock settings (Carrisosa *et al.*, 2021; da Silva *et al.*, 2022). For swine, modelling has examined *A.*

suum infection in Nordic countries (Nansen and Roepstorff, 1999) and population dynamics (Barnes, 1997).

Liver flukes

Overview

The liver flukes *Fasciola hepatica* and *F. gigantica* cause significant economic impact in livestock production, reducing productivity due to reduced milk and meat yields, condemned liver and greater susceptibility to other infections (Charlier *et al.*, 2007; Howell *et al.*, 2015). Although most fluke infections are subclinical in livestock, they are still of significant importance, for example reducing milk yields in dairy cattle by 8–15% (Howell *et al.*, 2015), and can have further effects on fertility (Charlier *et al.*, 2007).

Prevalence

Prevalence of *F. hepatica* is estimated at 48–76% in the UK (Salimi-Bejestani *et al.*, 2005; McCann *et al.*, 2010), 37% in Belgium, 50% in Germany and 61% in Spain (Mezo *et al.*, 2008; Bennema *et al.*, 2009; Kuerpick *et al.*, 2012). Climate change has increased reports of fasciolosis, due to wetter summers and warmer winters, expanding the range of the parasite (Fox *et al.*, 2011). Prevalence remains high despite the use of flukicide (Charlier *et al.*, 2014).

Risk factors

Climate and geographical factors play a key role in the risk of fasciolosis due to their effect on both the intermediate snail hosts and the survival and rate of development of the parasite on pasture (Howell *et al.*, 2015). A temperature of 10–15°C is required for both *Fasciola* larvae and snails to survive on pasture (Rapsch *et al.*, 2008). The presence of suitable snail habitats on pasture, the length of the grazing season and proportion of grazed grass in the diet, stocking rate and drinking water source are all additional risk factors (Bennema *et al.*, 2011; Charlier *et al.*, 2011); however, they vary between environments and farming systems (Charlier *et al.*, 2014).

Transmission

Transmission occurs when infective larvae are released from the intermediate freshwater snail host. The larvae then encyst on aquatic vegetation, which the mammalian definitive host ingests. Thus, transmission is heavily dependent on climatic conditions facilitative to the intermediate snail host and on season in temperate regions such as Europe (Khan *et al.*, 2017).

Control measures and biosecurity

Due to the indirect life cycle and intermediate host of *Fasciola* spp., it is important to reduce infection levels in snails as well as livestock (Knubben-Schweizer *et al.*, 2010). Triclobendazole has historically been the principal drug used in controlling fasciolosis but there are growing reports of resistance (Beesley *et al.*, 2023).

Availability of modelling data

Due to the seasonal and climate-dependent nature of *Fasciola* spp. transmission, a significant body of work has examined the role of climate, climate change (Fox *et al.*, 2011; Caminade *et al.*, 2015), the spatial dynamics of the fluke (Ducheyne *et al.*, 2015) and of the intermediate snail host in Switzerland (Rapsch *et al.*, 2008; Roessler *et al.*, 2022), Wales (Jones *et al.*, 2017), Ireland (Naranjo-Lucena *et al.*, 2018) and Sweden (Novobilský *et al.*, 2015), in addition to Europe-wide considerations (Caminade *et al.*, 2015; Cecconi *et al.*, 2016).

References

Abreu, R., Rodríguez-Álvarez, C., Lecuona, M., Castro-Hernández, B., González, J.C. *et al.* (2019) Prevalence and characteristics of methicillin-resistant staphylococci in goats on the island of Tenerife, Spain. *Acta Veterinaria Hungarica* 67(3), 317–326.

Ackermann, M. and Engels, M. (2006) Pro and contra IBR-eradication. *Veterinary Microbiology* 113(3–4), 293–302.

Aebi, M., Bodmer, M., Frey, J. and Pilo, P. (2012) Herd-specific strains of *Mycoplasma bovis* in outbreaks of Mycoplasmal mastitis and pneumonia. *Veterinary Microbiology* 157(3–4), 363–368.

Aebi, M., van den, B.H., Raemy, A., Steiner, A., Pilo, P. *et al.* (2015) *Mycoplasma bovis* infections in Swiss dairy cattle: A clinical investigation. *Acta Veterinaria Scandinavica* 57(1), 10.

Agnello, S., Chetta, M., Vicari, D., Mancuso, R., Manno, C. *et al.* (2012) Severe outbreaks of polyarthritis in kids caused by *Mycoplasma mycoides* subspecies *Capri* in Sicily. *American Journal of Veterinary Research* 44, 322–325.

Alarcon, P., Rushton, J. and Wieland, B. (2013) Cost of post-weaning multi-systemic wasting syndrome and porcine circovirus type-2 subclinical infection in England – an economic disease model. *Preventive Veterinary Medicine* 110(2), 88–102.

Alarcon, P., Velasova, M., Werling, D., Stärk, K.D.C., Chang, Y.-M. *et al.* (2011) Assessment and quantification of post-weaning multi-systemic wasting syndrome severity at farm level. *Preventive Veterinary Medicine* 98(1), 19–28.

Allan, G., Kennedy, S., McNeilly, F., Foster, J., Ellis, J. *et al.* (1999) Experimental reproduction of severe wasting disease by co-infection of pigs with porcine circovirus and porcine parvovirus. *Journal of Comparative Pathology* 121(1), 1–11.

Al-Mamun, M.A., Smith, R.L., Schukken, Y.H. and Gröhn, Y.T. (2016) Modeling of *Mycobacterium avium* subsp. *paratuberculosis* dynamics in a dairy herd: An individual based approach. *Journal of Theoretical Biology* 408, 105–117.

Al-Momani, W., Nicholas, R.A.J. and Abo-Shehada, M.N. (2008) Risk factors associated with *Mycoplasma agalactiae* infection of small ruminants in Northern Jordan. *Preventive Veterinary Medicine* 83(1), 1–10.

Alonso, C., Davies, P.R., Polson, D.D., Dee, S.A. and Lazarus, W.F. (2013a) Financial implications of installing air filtration systems to prevent PRRSV infection in large sow herds. *Preventive Veterinary Medicine* 111(3–4), 268–277.

Alonso, C., Murtaugh, M.P., Dee, S.A. and Davies, P.R. (2013b) Epidemiological study of air filtration systems for preventing PRRSV infection in large sow herds. *Preventive Veterinary Medicine* 112(1–2), 109–117.

Álvarez, L.G., Webb, C.R. and Holmes, M.A. (2011) A novel field-based approach to validate the use of network models for disease spread between dairy herds. *Epidemiology & Infection* 139(12), 1863–1874.

Aly, S.S. and Thurmond, M.C. (2005) Evaluation of *Mycobacterium avium* subsp *paratuberculosis* infection of dairy cows attributable to infection status of the dam. *Journal of the American Veterinary Medical Association* 227(3), 450–454.

Amass, S. and Baysinger, A. (2006) Swine disease transmission and prevention. In: Straw, B.E., Zimmerman, J.J., D'Allaire, S. and Taylor, D.J. (eds) *Diseases of Swine*, 9th edn. Blackwell, Oxford, pp. 1075–1098.

Amores, J., Sánchez, A., Gómez-Martín, A., Corrales, J.C., Contreras, A. *et al.* (2012) Surveillance of Mycoplasma agalactiae and Mycoplasma mycoides subsp. capri in dairy goat herds. *Small Ruminant Research* 102(1), 89–93.

Ampe, B., Goethals, K., Laevens, H. and Duchateau, L. (2012) Investigating clustering in interval-censored udder quarter infection times in dairy cows using a gamma frailty model. *Preventive Veterinary Medicine* 106(3–4), 251–257.

Andraud, M., Grasland, B., Durand, B., Cariolet, R., Jestin, A. *et al.* (2008a) Quantification of porcine circovirus type 2 (PCV-2) within- and between-pen transmission in pigs. *Veterinary Research* 39(5), 1.

Andraud, M., Grasland, B., Durand, B., Cariolet, R., Jestin, A. *et al.* (2008b) Estimation of PCV-2 transmission rate parameter derived from experimental data. *Épidémiologie et Santé Animale* (53), 19–28.

Andraud, M., Rose, N., Grasland, B., Pierre, J.S., Jestin, A. *et al.* (2009a) Influence of husbandry and control measures on porcine circovirus type 2 (PCV-2) dynamics within a farrow-to-finish pig farm: A modelling approach. *Preventive Veterinary Medicine* 92(1–2), 38–51.

Andraud, M., Rose, N., Grasland, B., Jestin, A. and Madec, F. (2009b) Modelling the within-herd dynamics of porcine circovirus type 2 (PCV-2) infection. Testing the influence of husbandry and prophylaxis measures. *Journées de La Recherche Porcine En France* 41, 185–190.

Andraud, M., Rose, N., Grasland, B., Jestin, A., Madec, F. *et al.* (2010) Modelling PCV-2 within-herd dynamics: A focus on pig-management practices. In: *Society for Veterinary Epidemiology and Preventive Medicine Proceedings*, Nantes, France, March 24–26, , pp. 127–134.

Andreopoulou, M., Chaligiannis, I., Sotiraki, S., Daugschies, A. and Bangoura, B. (2022) Prevalence and molecular detection of *Eimeria* species in different types of poultry in Greece and associated risk factors. *Parasitology Research* 121(7), 2051–2063.

Andreopoulou, M., Schares, G., Koethe, M., Chaligiannis, I., Maksimov, P. *et al.* (2023) Prevalence and molecular characterization of *Toxoplasma gondii* in different types of poultry in Greece, associated risk factors and co-existence with *Eimeria* spp. *Parasitology Research* 122(1), 97–111.

Animal Health Ireland (2021) Analysis of the economics of bohv-1 infection in Ireland. Teagasc. Available at: https://animalhealthireland.ie/assets/uploads/2021/04/AHI-IBR-Economic-Analysis-Report-2020.pdf?dl=1 (accessed 9 December 2024).

Antunes, A.C.L. and Jensen, D. (2019) Estimating time evolving cross-dependence of porcine reproduction and respiratory syndrome sero-prevalence in Danish swine herds. *Preventive Veterinary Medicine* 167, 85–89.

Arruda, A.G., Vilalta, C., Perez, A. and Morrison, R. (2017) Land altitude, slope, and coverage as risk factors for porcine reproductive and respiratory syndrome (PRRS) outbreaks in the United States. *PLoS One* 12(4), e0172638.

Arruda, A.G., Tousignant, S., Sanhueza, J., Vilalta, C., Poljak, Z. *et al.* (2019) Aerosol detection and transmission of porcine reproductive and respiratory syndrome virus (PRRSV): What is the evidence, and what are the knowledge gaps? *Viruses* 11(8), 712.

Assavacheep, P. and Rycroft, A. (2013) Survival of *Actinobacillus pleuropneumoniae* outside the pig. *Research in Veterinary Science* 94(1), 22–26.

Assavacheep, P. and Thanawongnuwech, R. (2022) Porcine respiratory disease complex: Dynamics of polymicrobial infections and management strategies after the introduction of the African swine fever. *Frontiers in Veterinary Science* 9, 1048861.

Assié, S., Bareille, N., Beaudeau, F. and Seegers, H. (2009) Management- and housing-related risk factors of respiratory disorders in non-weaned French Charolais calves. *Preventive Veterinary Medicine* 91(2–4), 218–225.

Assunção, P., Fe, C., Ramirez, A.S., Poveda, J.B. and Andrada, M. (2004) Serological study of contagious agalactia in herds of goats in the Canary Islands. *Veterinary Record* 154(22), 684–687.

Audigé, L., Doherr, M., Hauser, R. and Salman, M. (2001) Stochastic modelling as a tool for planning animal-health surveys and interpreting screening-test results. *Preventive Veterinary Medicine* 49(1–2), 1–17.

Ayele, W., Bartos, M., Svastova, P. and Pavlik, I. (2004) Distribution of *Mycobacterium avium* subsp. *paratuberculosis* in organs of naturally infected bull-calves and breeding bulls. *Veterinary Microbiology* 103(3–4), 209–217.

Baekbo, P. and Kristensen, C.S. (2015) PRRS control and eradication plans in Europe. Available at: www.pig333.com/articles/prrs-control-and-eradication-plans-in-europe_10668/: Pig333 (accessed 8 December 2024).

Baekbo, P., Kristensen, C. and Larsen, L. (2012) Porcine circovirus diseases: A review of PMWS. *Transboundary and Emerging Diseases* 59, 60–67.

Băieş, M.-H., Gherman, C., Boros, Z., Olah, D., Vlase, A.-M. *et al.* (2022) On the embryogenesis of *Ascaris suum* eggs during an in vitro experimental study. *Pathogens* 11(9), 1065.

Baker, J. and Frey, M. (1985) Bovine respiratory syncytial virus. *Veterinary Clinics of North America. Food Animal Practice* 1(2), 259–275.

Baker, J., Ames, T. and Markham, R. (1986) Seroepizootiologic study of bovine respiratory syncytial virus in a dairy herd. *American Journal of Veterinary Research* 47(2), 240–245.

Baptista, A.L., Rezende, A.L., de Almeida Fonseca, P., Massi, R.P., Nogueira, G.M. *et al.* (2017) Bovine respiratory disease complex associated mortality and morbidity rates in feedlot cattle from Southeastern Brazil. *Journal of Infection in Developing Countries* 11(10), 791–799.

Baranowski, E., Bergonier, D., Sagné, E., Hygonenq, M.-C., Ronsin, P. *et al.* (2014) Experimental infections with *Mycoplasma agalactiae* identify key factors involved in host-colonization. *PLoS One* 9(4), e93970.

Bargen, L.E. (2004) A system response to an outbreak of enzootic pneumonia in grow/finish pigs. *Canadian Veterinary Journal* 45(10), 856–859.

Barkema, H., Schukken, Y. and Zadoks, R. (2006) Invited review: The role of cow, pathogen, and treatment regimen in the therapeutic success of bovine *Staphylococcus aureus* mastitis. *Journal of Dairy Science* 89(6), 1877–1895.

Barkema, H.W., Green, M., Bradley, A.J. and Zadoks, R. (2009) Invited review: The role of contagious disease in udder health. *Journal of Dairy Science* 92(10), 4717–4729.

Barlow, J.W., White, L.J., Zadoks, R.N. and Schukken, Y.H. (2009) A mathematical model demonstrating indirect and overall effects of lactation therapy targeting subclinical mastitis in dairy herds. *Preventive Veterinary Medicine* 90(1–2), 31–42.

Barnes, E. (1997) Population dynamics of the parasitic stages of *Oesophagostomum dentatum* in pigs in single and trickle infections. *International Journal for Parasitology* 27(12), 1595–1604.

Barquero, N., Gomez-Lucia, E., Arjona, A., Toural, C., Las Heras, A. *et al.* (2013) Evolution of specific antibodies and proviral DNA in milk of small ruminants infected by small ruminant lentivirus. *Viruses* 5(10), 2614–2623.

Bastida, F. and Juste, R.A. (2011) Paratuberculosis control: A review with a focus on vaccination. *Journal of Immune Based Therapies and Vaccines* 9, 1–17.

Beaunée, G., Ezanno, P., Joly, A., Nicolas, P. and Vergu, E. (2017) Inference in a metapopulation model of paratuberculosis in cattle via a composite-likelihood approximation. In: *Modelling in Animal Health Conference*, p. 69.

Beesley, N.J., Cwiklinski, K., Allen, K., Hoyle, R.C., Spithill, T.W. *et al.* (2023) A major locus confers triclabendazole resistance in *Fasciola hepatica* and shows dominant inheritance. *PLoS Pathogens* 19(1), e1011081.

Benavides, B., Casal, J., Diéguez, J., Yus, E., Moya, S.J. *et al.* (2020) Development of a quantitative risk assessment of bovine viral diarrhea virus and bovine herpesvirus-1 introduction in dairy cattle herds to improve biosecurity. *Journal of Dairy Science* 103(7), 6454–6472.

Benavides, B., Casal, J., Diéguez, J., Yus, E., Moya, S. *et al.* (2021) Quantitative risk assessment of introduction of BVDV and BoHV-1 through indirect contacts based on implemented biosecurity measures in dairy farms of Spain. *Preventive Veterinary Medicine* 188, 105263.

Benedictus, G., Verhoeff, J., Schukken, Y. and Hesselink, J. (2000) Dutch paratuberculosis programme: History, principles and development. *Veterinary Microbiology* 77(3–4), 399–413.

Bennema, S., Vercruysse, J., Claerebout, E., Schnieder, T., Strube, C. *et al.* (2009) The use of bulk-tank milk ELISAs to assess the spatial distribution of *Fasciola hepatica*, *Ostertagia ostertagi* and *Dictyocaulus viviparus* in dairy cattle in Flanders (Belgium). *Veterinary Parasitology* 165(1–2), 51–57.

Bennema, S., Ducheyne, E., Vercruysse, J., Claerebout, E., Hendrickx, G. *et al.* (2011) Relative importance of management, meteorological and environmental factors in the spatial distribution of *Fasciola hepatica* in dairy cattle in a temperate climate zone. *International Journal for Parasitology* 41(2), 225–233.

Ben Romdhane, R., Beaunée, G., Camanes, G., Guatteo, R., Fourichon, C. *et al.* (2017) Which phenotypic traits of resistance should be improved in cattle to control paratuberculosis dynamics in a dairy herd: A modelling approach. *Veterinary Research* 48(1), 62.

Bergonier, D., Berthelot, X. and Poumarat, F. (1997) Contagious agalactia of small ruminants: Current knowledge concerning epidemiology, diagnosis and control. *Revue Scientifique et Technique (International Office of Epizootics)* 16(3), 848–873.

Bernal, J.M.C., Fernández, A., Arnal, J.L., Baselga, C., Benito Zuñiga, A. *et al.* (2023) Cluster analysis of bovine respiratory disease (BRD)-associated pathogens shows the existence of two epidemiological patterns in BRD outbreaks. *Veterinary Microbiology* 280, 109701.

Besier, R., Kahn, L., Sargison, N. and Wyk, J.A. (2016) The pathophysiology, ecology and epidemiology of *Haemonchus contortus* infection in small ruminants. *Advances in Parasitology* 93, 95–143.

Biddle, M.K., Fox, L.K., Evans, M.A. and Gay, C.C. (2005) Pulsed-field gel electrophoresis patterns of *Mycoplasma* isolates from various body sites in dairy cattle with *Mycoplasma* mastitis. *Journal of the American Veterinary Medical Association* 227(3), 455–459.

Bidokhti, M.R., Tråvén, M., Fall, N., Emanuelson, U. and Alenius, S. (2009) Reduced likelihood of bovine coronavirus and bovine respiratory syncytial virus infection on organic compared to conventional dairy farms. *Veterinary Journal* 182(3), 436–440.

Bidokhti, M.R., Tråvén, M., Ohlson, A., Zarnegar, B., Baule, C. *et al.* (2012) Phylogenetic analysis of bovine respiratory syncytial viruses from recent outbreaks in feedlot and dairy cattle herds. *Archives of Virology* 157(4), 601–607.

Biebaut, E., Chantziaras, I., Boyen, F., Devriendt, B., Haesebrouck, F. *et al.* (2022) Influence of parity and reproductive stage on the prevalence of *Mycoplasma hyopneumoniae* in breeding animals in Belgian farrow-to-finish pig herds. *Porcine Health Management* 8(1), 26.

Biemans, F., Romdhane, R.B., Gontier, P., Fourichon, C., Ramsbottom, G. *et al.* (2021) Modelling transmission and control of *Mycobacterium avium* subspecies *paratuberculosis* within Irish dairy herds with compact spring calving. *Preventive Veterinary Medicine* 186, 105228.

Biemans, F., Tratalos, J., Arnoux, S., Ramsbottom, G., More, S.J. *et al.* (2022) Modelling transmission of *Mycobacterium avium* subspecies *paratuberculosis* between Irish dairy cattle herds. *Veterinary Research* 53(1), 45.

Bitsch, V. (1973) Infectious bovine rhinotracheitis virus infection in bulls, with special reference to preputial infection. *Applied Microbiology* 26(3), 337–343.

Bitsch, V., Hansen, K.-E. and Rønsholt, L. (2000) Experiences from the Danish programme for eradication of bovine virus diarrhoea (BVD) 1994–1998 with special reference to legislation and causes of infection. *Veterinary Microbiology* 77(1–2), 137–143.

Bitsouni, V., Lycett, S., Opriessnig, T. and Doeschl-Wilson, A. (2019) Predicting vaccine effectiveness in livestock populations: A theoretical framework applied to PRRS virus infections in pigs. *PLoS One* 14(8), e0220738.

Blacklaws, B.A., Berriatua, E., Torsteinsdottir, S., Watt, N.J., Andres, D. *et al.* (2004) Transmission of small ruminant lentiviruses. *Veterinary Microbiology* 101(3), 199–208.

Blakebrough-Hall, C., McMeniman, J.P. and González, L.A. (2020) An evaluation of the economic effects of bovine respiratory disease on animal performance, carcass traits, and economic outcomes in feedlot cattle defined using four BRD diagnosis methods. *Journal of Animal Science* 98(2), skaa005.

Blake, D.P., Knox, J., Dehaeck, B., Huntington, B., Rathinam, T. *et al.* (2020) Re-calculating the cost of coccidiosis in chickens. *Veterinary Research* 51(1), 115.

Blickenstorfer, S., Schwermer, H., Engels, M., Reist, M., Doherr, M.G. *et al.* (2011) Using scenario tree modelling for targeted herd sampling to substantiate freedom from disease. *BMC Veterinary Research* 7(1), 1–9.

Bokma, J., Stuyvaert, S. and Pardon, B. (2022) Comparison and optimisation of screening cutoff values for *Mycoplasma bovis* antibody ELISAs using serum from youngstock. *Veterinary Record* 191(9), e2179.

Bolea, R., Monleón, E., Carrasco, L., Vargas, A., de Andrés, D. *et al.* (2006) Maedi-visna virus infection of ovine mammary epithelial cells. *Veterinary Research* 37(1), 133–144.

Borjigin, L., Shimazu, T., Katayama, Y., Li, M., Satoh, T. *et al.* (2016) Immunogenic properties of landrace pigs selected for resistance to *Mycoplasma pneumonia* of swine. *Animal Science Journal* 87(3), 321–329.

Bøtner, A. and Belsham, G.J. (2012) Virus survival in slurry: Analysis of the stability of foot-and-mouth disease, classical swine fever, bovine viral diarrhoea and swine influenza viruses. *Veterinary Microbiology* 157(1–2), 41–49.

Boulbria, G., Brilland, S., Teixeira-Costa, C., Brissonnier, M., Charles, M. *et al.* (2021) Effectiveness of two intramuscular combined vaccines for the control of *Mycoplasma hyopneumoniae* and porcine circovirus type 2 in growing pigs: A randomized field trial. *Porcine Health Management* 7(1), 41.

Brock, J., Lange, M., Tratalos, J.A., More, S.J., Guelbenzu-Gonzalo, M. *et al.* (2021) Thulke H-H: A large-scale epidemiological model of bohv-1 spread in the Irish cattle population to support decision-making in conformity with the European animal health law. *Preventive Veterinary Medicine* 192, 105375.

Broughan, J., Judge, J., Ely, E., Delahay, R.J., Wilson, G. *et al.* (2016) A review of risk factors for bovine tuberculosis infection in cattle in the UK and Ireland. *Epidemiology and Infection* 144(14), 2899–2926.

Browne, C., Loeffler, A., Holt, H., Chang, Y., Lloyd, D. *et al.* (2017) Low temperature and dust favour in vitro survival of *Mycoplasma hyopneumoniae*: Time to revisit indirect transmission in pig housing. *Letters in Applied Microbiology* 64(1), 2–7.

Burke, J.M. and Miller, J.E. (2020) Sustainable approaches to parasite control in ruminant livestock. *Veterinary Clinics of North America. Food Animal Practice* 36(1), 89–107.

Busin, V., Kenyon, F., Parkin, T., McBean, D., Laing, N. *et al.* (2014) Production impact of a targeted selective treatment system based on liveweight gain in a commercial flock. *Veterinary Journal* 200(2), 248–252.

Butler, J.A., Sickles, S.A., Johanns, C.J. and Rosenbusch, R.F. (2000) Pasteurization of discard mycoplasma mastitic milk used to feed calves: Thermal effects on various mycoplasma. *Journal of Dairy Science* 83(10), 2285–2288.

Callan, R.J. and Garry, F.B. (2002) Biosecurity and bovine respiratory disease. *Veterinary Clinics of North America. Food Animal Practice* 18(1), 57–77.
Calsamiglia, M. and Pijoan, C. (2000) Colonisation state and colostral immunity to *Mycoplasma hyopneumoniae* of different parity sows. *Veterinary Record* 146(18), 530–532.
Camanes, G., Joly, A., Fourichon, C., Ben Romdhane, R. and Ezanno, P. (2018) Control measures to prevent the increase of paratuberculosis prevalence in dairy cattle herds: An individual-based modelling approach. *Veterinary Research* 49(1), 60.
Caminade, C., Dijk, J., Baylis, M. and Williams, D. (2015) Modelling recent and future climatic suitability for fasciolosis in Europe. *Geospatial Health* 9(2), 301–308.
Campler, M.R., Cheng, T.-Y., Schroeder, D.C., Yang, M., Mor, S.K. *et al.* (2022) A longitudinal study on PRRSV detection in swine herds with different demographics and PRRSV management strategies. *Transboundary and Emerging Diseases* 69(4), e1005–e1014.
Campos, B., Pickering, A.C., Rocha, L.S., Aguilar, A.P., Fabres-Klein, M.H. *et al.* (2022) Diversity and pathogenesis of *Staphylococcus aureus* from bovine mastitis: Current understanding and future perspectives. *BMC Veterinary Research* 18(1), 1–16.
Carrisosa, M., Jin, S., McCrea, B.A., Macklin, K.S., Dormitorio, T. *et al.* (2021) Prevalence of select intestinal parasites in Alabama backyard poultry flocks. *Animals* 11(4), 939.
Casey-Bryars, M., Tratalos, J.A., Graham, D.A., Guelbenzu-Gonzalo, M.P., Barrett, D. *et al.* (2022) Risk factors for detection of bovine viral diarrhoea virus in low-risk herds during the latter stages of Ireland's eradication programme. *Preventive Veterinary Medicine* 201, 105607.
Cecconi, L., Grisotto, L., Catelan, D., Lagazio, C., Berrocal, V. *et al.* (2016) Preferential sampling and Bayesian geostatistics: Statistical modeling and examples. *Statistical Methods in Medical Research* 25(4), 1224–1243.
Chamorro, M.F., Passler, T., Givens, M.D., Edmondson, M.A., Wolfe, D.F. *et al.* (2011) Evaluation of transmission of bovine viral diarrhea virus (BVDV) between persistently infected and naive cattle by the horn fly (*Haematobia irritans*). *Veterinary Research Communications* 35(2), 123–129.
Chapman, H.D. (2014) Milestones in avian coccidiosis research: A review. *Poultry Science* 93(3), 501–511.
Charlier, J., Duchateau, L., Claerebout, E., Williams, D. and Vercruysse, J. (2007) Associations between anti-*Fasciola hepatica* antibody levels in bulk-tank milk samples and production parameters in dairy herds. *Preventive Veterinary Medicine* 78(1), 57–66.
Charlier, J., Bennema, S.C., Caron, Y., Counotte, M., Ducheyne, E. *et al.* (2011) Towards assessing fine-scale indicators for the spatial transmission risk of *Fasciola hepatica* in cattle. *Geospatial Health* 5(2), 239–245.
Charlier, J., Vercruysse, J., Morgan, E., Dijk, J. and Williams, D. (2014) Recent advances in the diagnosis, impact on production and prediction of *Fasciola hepatica* in cattle. *Parasitology* 141(3), 326–335.
Charlier, J., Thamsborg, S.M., Bartley, D.J., Skuce, P.J., Kenyon, F. *et al.* (2018) Mind the gaps in research on the control of gastrointestinal nematodes of farmed ruminants and pigs. *Transboundary and Emerging Diseases* 65, 217–234.
Charlier, J., Rinaldi, L., Musella, V., Ploeger, H.W., Chartier, C. *et al.* (2020) Initial assessment of the economic burden of major parasitic helminth infections to the ruminant livestock industry in Europe. *Preventive Veterinary Medicine* 182, 105103.
Chauve, C. (1998) The poultry red mite *Dermanyssus gallinae* (De Geer, 1778): Current situation and future prospects for control. *Veterinary Parasitology* 79(3), 239–245.
Chazel, M., Tardy, F., Le Grand, D., Calavas, D. and Poumarat, F. (2010) Mycoplasmoses of ruminants in France: Recent data from the national surveillance network. *BMC Veterinary Research* 6(1), 1–8.
Chiodini, R.J. and Hermon-Taylor, J. (1993) The thermal resistance of *Mycobacterium paratuberculosis* in raw milk under conditions simulating pasteurization. *Journal of Veterinary Diagnostic Investigation* 5(4), 629–631.
Chiu, L.J.V., Tauer, L.W., Al-Mamun, M.A., Kaniyamattam, K., Smith, R.L. *et al.* (2018) An agent-based model evaluation of economic control strategies for paratuberculosis in a dairy herd. *Journal of Dairy Science* 101(7), 6443–6454.
Christensen, C., Barnes, E., Nansen, P., Roepstorff, A. and Slotved, H.-C. (1995) Experimental *Oesophagostomum dentatum* infection in the pig: Worm populations resulting from single infections with three doses of larvae. *International Journal for Parasitology* 25(12), 1491–1498.
Cillara, G., Manca, M.G., Longheu, C. and Tola, S. (2015) Discrimination between *Mycoplasma mycoides* subsp. *capri* and *Mycoplasma capricolum* subsp. *capricolum* using PCR-RFLP and PCR. *Veterinary Journal* 205(3), 421–423.

Cirone, F., Padalino, B., Tullio, D., Capozza, P., Lo Surdo, M. *et al.* (2019) Prevalence of pathogens related to bovine respiratory disease before and after transportation in beef steers: Preliminary results. *Animals* 9(12), 1093.

Ciszewski, D.K., Baker, J.C., Slocombe, R.F., Reindel, J.F., Haines, D.M. *et al.* (1991) Experimental reproduction of respiratory tract disease with bovine respiratory syncytial virus. *Veterinary Microbiology* 28(1), 39–60.

Claerebout, E. and Geldhof, P. (2020) Helminth vaccines in ruminants: From development to application. *Veterinary Clinics of North America. Food Animal Practice* 36(1), 159–171.

Clilverd, H., Martín-Valls, G., Li, Y., Martín, M., Cortey, M. *et al.* (2023) Infection dynamics, transmission, and evolution after an outbreak of porcine reproductive and respiratory syndrome virus. *Frontiers in Microbiology* 14, 1109881.

Cobb, S.P., Pharo, H., Stone, M., Groenendaal, H. and Zagmutt, F.J. (2015) Quantitative risk assessment of the likelihood of introducing porcine reproductive and respiratory syndrome virus into New Zealand through the importation of pig meat. *Revue Scientifique et Technique de l'OIE* 34(3), 961–975.

Coles, G., Jackson, F., Pomroy, W., Prichard, R., von Samson-Himmelstjerna, G. *et al.* (2006) The detection of anthelmintic resistance in nematodes of veterinary importance. *Veterinary Parasitology* 136(3–4), 167–185.

Corrales, J., Esnal, A., de la Fe, C., Sánchez, A., Assunçao, P. *et al.* (2007) Contagious agalactia in small ruminants. *Small Ruminant Research* 68(1–2), 154–166.

Cortey, M., Pileri, E., Sibila, M., Pujols, J., Balasch, M. *et al.* (2011) Genotypic shift of porcine circovirus type 2 from PCV-2a to PCV-2b in Spain from 1985 to 2008. *Veterinary Journal* 187(3), 363–368.

Courcoul, A. and Ezanno, P. (2010) Modelling the spread of bovine viral diarrhoea virus (BVDV) in a managed metapopulation of cattle herds. *Veterinary Microbiology* 142(1–2), 119–128.

Cowled, B.D., Stevenson, M.A. and Madin, B. (2016) An assessment of the association between soil ph and ovine Johne's disease using Australian abattoir surveillance data. *Preventive Veterinary Medicine* 126, 208–219.

Cowled, B.D., Sergeant, E.S., Leslie, E.E., Crosbie, A., Burroughs, A. *et al.* (2022) Use of scenario tree modelling to plan freedom from infection surveillance: *Mycoplasma bovis* in New Zealand. *Preventive Veterinary Medicine* 198, 105523.

Cusack, P. (2023) Evaluation of practices used to reduce the incidence of bovine respiratory disease in Australian feedlots (to November 2021). *Australian Veterinary Journal* 101(6), 230–247.

Damman, A., Viet, A.-F., Arnoux, S., Guerrier-Chatellet, M.-C., Petit, E. *et al.* (2015) Modelling the spread of bovine viral diarrhea virus (BVDV) in a beef cattle herd and its impact on herd productivity. *Veterinary Research* 46, 1–14.

Dank, M., Holzhauer, M., Veldhuis, A. and Frankena, K. (2015) Association between *Dictyocaulus viviparus* status and milk production parameters in Dutch dairy herds. *Journal of Dairy Science* 98(11), 7741–7747.

da Silva, J.T. d, Alvares, F.B.V., Oliveira, C.S. d M., Sousa, L.C. d, Lima, B.A. *et al.* (2022) Prevalence of endoparasites by microscopic analysis in free-range chickens in a Brazilian semiarid region. *Revista Brasileira de Parasitologia Veterinaria* 31(4), e010722.

David, K., Vergara, H., Eger, S., Klassen, A., Schwödiauer, P. *et al.* (2021) Beziehungen zwischen managementfaktoren und bestandsstatus bezüglich infektion mit dem virus des porzinen reproduktiven und respiratorischen syndroms. *Tierärztliche Praxis Ausgabe G: Großtiere/Nutztiere* 49(1), 30–39.

de Miguel, R., Arrieta, M., Rodríguez-Largo, A., Echeverría, I., Resendiz, R. *et al.* (2021) Worldwide prevalence of small ruminant lentiviruses in sheep: A systematic review and meta-analysis. *Animals* 11(3), 784.

De la Fe, C., Gutiérrez, A., Poveda, J.B., Assunção, P., Ramírez, A.S. *et al.* (2007) First isolation of *Mycoplasma capricolum* subsp. *capricolum*, one of the causal agents of caprine contagious Agalactia, on the island of Lanzarote (Spain). *Veterinary Journal* 173(2), 440–442.

De Paz, X., Vega, D., Duran, C. and Angulo, J. (2015) PRRS prevalence in Europe: Perception of the pig veterinary practitioners. *European Society of Porcine Health Management* 7, 155.

Decru, E., Mul, M., Nisbet, A.J., Vargas Navarro, A.H., Chiron, G. *et al.* (2020) Possibilities for IPM strategies in European laying hen farms for improved control of the poultry red mite (*Dermanyssus gallinae*): Details and state of affairs. *Frontiers in Veterinary Science* 7, 565866.

Dee, S., Torremorell, M., Thompson, B., Deen, J. and Pijoan, C. (2005) An evaluation of thermo-assisted drying and decontamination for the elimination of porcine reproductive and respiratory syndrome virus from contaminated livestock transport vehicles. *Canadian Journal of Veterinary Research* 69(1), 58–63.

Dee, S., Otake, S., Oliveira, S. and Deen, J. (2009) Evidence of long distance airborne transport of porcine reproductive and respiratory syndrome virus and *Mycoplasma hyopneumoniae*. *Veterinary Research* 40(4), 39.

Dee, S.A., Deen, J., Cano, J.P., Batista, L. and Pijoan, C. (2006) Further evaluation of alternative air-filtration systems for reducing the transmission of porcine reproductive and respiratory syndrome virus by aerosol. *Canadian Journal of Veterinary Research* 70(3), 168–175.

Dee, S.A., Bauermann, F.V., Niederwerder, M.C., Singrey, A., Clement, T. *et al.* (2018) Survival of viral pathogens in animal feed ingredients under transboundary shipping models. *PloS One* 13(3), e0194509.

Deeney, A.S., Collins, R. and Ridley, A.M. (2021) Identification of *Mycoplasma* species and related organisms from ruminants in England and Wales during 2005–2019. *BMC Veterinary Research* 17(1), 1–12.

Dee, S.A., Niederwerder, M.C., Edler, R., Hanson, D., Singrey, A. *et al.* (2021) An evaluation of additives for mitigating the risk of virus-contaminated feed using an ice-block challenge model. *Transboundary and Emerging Diseases* 68(2), 833–845.

Delabouglise, A., James, A., Valarcher, J.-F., Hagglünd, S., Raboisson, D. *et al.* (2017) Linking disease epidemiology and livestock productivity: The case of bovine respiratory disease in France. *PLoS One* 12(12), e0189090.

Deluyker, H., Oye, S. and Boucher, J. (2005) Factors affecting cure and somatic cell count after pirlimycin treatment of subclinical mastitis in lactating cows. *Journal of Dairy Science* 88(2), 604–614.

Desrosiers, R. (2011) Transmission of swine pathogens: Different means, different needs. *Animal Health Research Reviews* 12(1), 1–13.

Desrosiers, R., Boutin, R. and Broes, A. (2004) Persistence of antibodies after natural infection with swine influenza virus and epidemiology of the infection in a herd previously considered influenza-negative. *Journal of Swine Health and Production* 12(2), 78–81.

Dinsmore, R.P. (2002) Biosecurity for mammary diseases in dairy cattle. *Veterinary Clinics of North America* 18(1), 115–131.

Dione, M., Masembe, C., Akol, J., Amia, W., Kungu, J. *et al.* (2018) The importance of on-farm biosecurity: Sero-prevalence and risk factors of bacterial and viral pathogens in smallholder pig systems in Uganda. *Acta Tropica* 187, 214–221.

Divers, T.J. (2007) Respiratory diseases. In: Divers, T. and Peek, S. (eds) *Rebhun's Diseases of Dairy Cattle*, 2nd edn. Elsevier, Philadelphia, pp. 79–129.

Doré, E., Paré, J., Côté, G., Buczinski, S., Labrecque, O. *et al.* (2012) Risk factors associated with transmission of *Mycobacterium avium* subsp. *paratuberculosis* to calves within dairy herd: A systematic review. *Journal of Veterinary Internal Medicine* 26(1), 32–45.

Dos Santos, L.F., Sreevatsan, S., Torremorell, M., Moreira, M.A.S., Sibila, M. *et al.* (2015) Genotype distribution of *Mycoplasma hyopneumoniae* in swine herds from different geographical regions. *Veterinary Microbiology* 175(2–4), 374–381.

Ducheyne, E., Charlier, J., Vercruysse, J., Rinaldi, L., Biggeri, A. *et al.* (2015) Modelling the spatial distribution of *Fasciola hepatica* in dairy cattle in Europe. *Geospatial Health* 9(2), 261–270.

Dupont, K., Nielsen, E.O., Baekbo, P. and Larsen, L.E. (2008) Genomic analysis of PCV2 isolates from Danish archives and a current PMWS case–control study supports a shift in genotypes with time. *Veterinary Microbiology* 128(1–2), 56–64.

EFSA AHAW Panel (EFSA Panel on Animal Health and Welfare), Nielsen, S.S., Bicout, D.J., Calistri, P., Canali, E. *et al.* (2021) Scientific opinion on the assessment of animal diseases caused by bacteria resistant to antimicrobials: Cattle. *EFSA Journal* 19(12), 89. DOI: 10.2903/j.efsa.2021.6955.

Eisenberg, S.W., Nielen, M. and Koets, A.P. (2012) Within-farm transmission of bovine paratuberculosis: Recent developments. *Veterinary Quarterly* 32(1), 31–35.

Ellis, J.A., Gow, S.P. and Goji, N. (2010) Response to experimentally induced infection with bovine respiratory syncytial virus following intranasal vaccination of seropositive and seronegative calves. *Journal of the American Veterinary Medical Association* 236(9), 991–999.

Ellis, J.A., Philibert, H., West, K., Clark, E., Martin, K. *et al.* (1996) Fatal pneumonia in adult dairy cattle associated with active infection with bovine respiratory syncytial virus. *Canadian Veterinary Journal* 37(2), 103–105.

Elvander, M. (1996) Severe respiratory disease in dairy cows caused by infection with bovine respiratory syncytial virus. *Veterinary Record* 138(5), 101–105.

Ersbøll, A.K., Ersbøll, B.K., Houe, H., Alban, L. and Kjeldsen, A. (2010) Spatial modelling of the between-herd infection dynamics of bovine virus diarrhoea virus (BVDV) in dairy herds in Denmark. *Preventive Veterinary Medicine* 97(2), 83–89.

Eslahi, A.V., KarimiPourSaryazdi, A., Olfatifar, M., Carvalho, L.M.M., Foroutan, M. *et al.* (2022) Global prevalence of trichinella in pigs: A systematic review and meta-analysis. *Veterinary Medicine and Science* 8(6), 2466–2481.

Eterpi, M., McDonnell, G. and Thomas, V. (2011) Decontamination efficacy against *Mycoplasma*. *Letters in Applied Microbiology* 52(2), 150–155.

European Commission (2020) Commission Implementing Regulation (EU) 2020/2002 of 7 December 2020 laying down rules for the application of Regulation (EU) 2016/429 of the European Parliament and of the Council with regard to Union notification and Union reporting of listed diseases, to formats and procedures for submission and reporting of Union surveillance programmes and of eradication programmes and for application for recognition of disease-free status, and to the computerised information system. Available at: https://eur-lex.europa.eu/eli/reg_impl/2020/2002/oj (accessed 12 December 2024).

Evans, C.M., Medley, G.F., Creasey, S.J. and Green, L.E. (2010) A stochastic mathematical model of the within-herd transmission dynamics of porcine reproductive and respiratory syndrome virus (PRRSV): Fade-out and persistence. *Preventive Veterinary Medicine* 93(4), 248–257.

Exel, C.E., Halasa, T., Koop, G., Steeneveld, W., Lam, T.J. *et al.* (2022) A stochastic modelling approach to determine the effect of diverse *Staphylococcus aureus* strains on the economic and epidemiological outcomes of mastitis intervention strategies in dairy cattle. *Preventive Veterinary Medicine* 199, 105566.

Exel, C.E., Gerritsen, K., Spaninks, M., Duim, B., Koop, G. *et al.* (2023) Association of *Staphylococcus aureus* genotypes with milk or colonization of extramammary sites in Dutch dairy cattle indicates strain variation in reservoirs for intramammary infections. *Research in Veterinary Science* 154, 138–144.

Eysker, M., Bakker, N., Kooyman, F., Linden, D., Schrama, C. *et al.* (2005) Consequences of the unusually warm and dry summer of 2003 in the Netherlands: Poor development of free living stages, normal survival of infective larvae and long survival of adult gastrointestinal nematodes of sheep. *Veterinary Parasitology* 133(4), 313–321.

Ezanno, P., Fourichon, C., Viet, A.-F. and Seegers, H. (2007) Sensitivity analysis to identify key-parameters in modelling the spread of bovine viral diarrhoea virus in a dairy herd. *Preventive Veterinary Medicine* 80(1), 49–64.

Ezanno, P., Arnoux, S., Joly, A. and Vermesse, R. (2022) Rewiring cattle trade movements helps to control bovine paratuberculosis at a regional scale. *Preventive Veterinary Medicine* 198, 105529.

Fablet, C., Marois-Créhan, C., Grasland, B., Simon, G. and Rose, N. (2016) Factors associated with herd-level PRRSV infection and age-time to seroconversion in farrow-to-finish herds. *Veterinary Microbiology* 192, 10–20.

Fagundes, H., Barchesi, L., Nader Filho, A., Ferreira, L.M. and Oliveira, C.A.F. (2010) Occurrence of *Staphylococcus aureus* in raw milk produced in dairy farms in São Paulo state, Brazil. *Brazilian Journal of Microbiology* 41(2), 376–380.

Fecteau, M.-E. and Whitlock, R.H. (2011) Treatment and chemoprophylaxis for paratuberculosis. *Veterinary Clinics of North America. Food Animal Practice* 27(3), 547–557.

Fecteau, M.-E. (2018) Paratuberculosis in cattle. *Veterinary Clinics of North America* 34(1), 209–222.

Fernández, G., Barreal, M., Pombo, M., Ginzo-Villamayor, M., González-Manteiga, W. *et al.* (2013) Comparison of the epidemiological behavior of mastitis pathogens by applying time-series analysis in results of milk samples submitted for microbiological examination. *Veterinary Research Communications* 37(4), 259–267.

Ferraz, M., Almeida, H., Storino, G., Sonálio, K., Souza, M. *et al.* (2020) Lung consolidation caused by *Mycoplasma hyopneumoniae* has a negative effect on productive performance and economic revenue in finishing pigs. *Preventive Veterinary Medicine* 182, 105091.

Field, N.L., Mee, J.F. and McAloon, C.G. (2023) Evaluation of environmental sampling for detection of *Mycobacterium avium* subspecies *paratuberculosis* in the pre-weaned calf area and calving area of infected dairy farms enrolled in a voluntary Johne's disease control programme. *Animals* 13(4), 669.

Filioussis, G., Petridou, E., Giadinis, N.D. and Kritas, S.K. (2014) *In vitro* susceptibilities of caprine *Mycoplasma agalactiae* field isolates to six antimicrobial agents using the E test methodology. *Veterinary Journal* 202(3), 654–656.

Filipe, J., Kyriazakis, I., McFarland, C. and Morgan, E. (2023) Novel epidemiological model of gastrointestinal nematode infection to assess grazing cattle resilience by integrating host growth, parasite, grass and environmental dynamics. *International Journal for Parasitology* 53(3), 133–155.
Fluit, A. (2012) Livestock-associated *Staphylococcus aureus*. *Clinical Microbiology and Infection* 18(8), 735–744.
Foddai, A., Enøe, C., Krogh, K., Stockmarr, A. and Halasa, T. (2014) Stochastic simulation modeling to determine time to detect bovine viral diarrhea antibodies in bulk tank milk. *Preventive Veterinary Medicine* 117(1), 149–159.
Foster, A.P., Naylor, R.D., Howie, N.M., Nicholas, R.A.J. and Ayling, R.D. (2009) *Mycoplasma bovis* and otitis in dairy calves in the United Kingdom. *Veterinary Journal* 179(3), 455–457.
Fountain, J., Brookes, V., Kirkeby, C., Manyweathers, J., Maru, Y. *et al*. (2022a) Simulating the economic and non-economic impact of on-farm biosecurity for BVDV in Australian beef production: The impact of on-farm biosecurity for BVDV in Australian beef production. In: *2nd Australian Biosecurity Symposium*. Available at: www.biosym.com.au/2nd-australian-biosecurity-symposium/ (accessed 13 December 2024).
Fountain, J., Brookes, V., Kirkeby, C., Manyweathers, J., Maru, Y. *et al*. (2022b) One size does not fit all: the impact of on-farm biosecurity for BVDV in Australian beef production. *Preventive Veterinary Medicine* 208, 105758.
Fox, L., Kirk, J. and Britten, A. (2005) *Mycoplasma* mastitis: A review of transmission and control. *Journal of Veterinary Medicine. B, Infectious Diseases and Veterinary Public Health* 52(4), 153–160.
Fox, L.K. (2012) *Mycoplasma* mastitis: Causes, transmission, and control. *Veterinary Clinics of North America. Food Animal Practice* 28(2), 225–237.
Fox, N.J., White, P.C., McClean, C.J., Marion, G., Evans, A. *et al*. (2011) Predicting impacts of climate change on *Fasciola hepatica* risk. *PLoS One* 6(1), e16126.
Fraile, L., Fernández, N., Pena, R.N., Balasch, S., Castellà, G. *et al*. (2020) A probabilistic poisson-based model to detect PRRSV recirculation using sow production records. *Preventive Veterinary Medicine* 177, 104948.
Franzo, G., Faustini, G., Legnardi, M., Cecchinato, M., Drigo, M. *et al*. (2022) Phylodynamic and phylogeographic reconstruction of porcine reproductive and respiratory syndrome virus (PRRSV) in Europe: Patterns and determinants. *Transboundary and Emerging Diseases* 69(5), e2175–e2184.
Frey, J. (1995) Virulence in *Actinobacillus pleuropneumoniae* and RTX toxins. *Trends in Microbiology* 3(7), 257–261.
Frössling, J., Agren, E.C.C., Eliasson-Selling, L. and Lewerin, S.S. (2009) Probability of freedom from disease after the first detection and eradication of PRRS in Sweden: Scenario-tree modelling of the surveillance system. *Preventive Veterinary Medicine* 91(2–4), 137–145.
Fujimoto, Y., Ito, H., Higuchi, H., Ohno, H. and Makita, K. (2020) A case–control study of herd- and cow-level risk factors associated with an outbreak of *Mycoplasma* mastitis in Nemuro, Japan. *Preventive Veterinary Medicine* 177, 104946.
Fulton, R.W., Whitley, E.M., Johnson, B.J., Ridpath, J.F., Kapil, S. *et al*. (2009) Prevalence of bovine viral diarrhea virus (BVDV) in persistently infected cattle and BVDV subtypes in affected cattle in beef herds in South Central United States. *Canadian Journal of Veterinary Research* 73(4), 283–291.
Fusco, M., Corona, L., Onni, T., Marras, E., Longheu, C. *et al*. (2007) Development of a sensitive and specific enzyme-linked immunosorbent assay based on recombinant antigens for rapid detection of antibodies against *Mycoplasma agalactiae* in sheep. *Clinical and Vaccine Immunology* 14(4), 420–425.
Gagnon, C.A., Tremblay, D., Tijssen, P., Venne, M.H., Houde, A. *et al*. (2007) The emergence of porcine circovirus 2b genotype (PCV-2b) in swine in Canada. *Canadian Veterinary Journal* 48(8), 811–819.
Galli, F., Friker, B., Bearth, A. and Dürr, S. (2022) Direct and indirect pathways for the spread of African swine fever and other porcine infectious diseases: An application of the mental models approach. *Transboundary and Emerging Diseases* 69(5), e2602–e2616.
Galvis, J.A., Corzo, C.A. and Machado, G. (2022a) Modelling and assessing additional transmission routes for porcine reproductive and respiratory syndrome virus: Vehicle movements and feed ingredients. *Transboundary and Emerging Diseases* 69(5), e1549–e1560.
Galvis, J.A., Corzo, C.A., Prada, J.M. and Machado, G. (2022b) Modelling the transmission and vaccination strategy for porcine reproductive and respiratory syndrome virus. *Transboundary and Emerging Diseases* 69(2), 485–500.

Garcia, A.B. and Shalloo, L. (2015) Invited review: The economic impact and control of paratuberculosis in cattle. *Journal of Dairy Science* 98(8), 5019–5039.

Garcia-Morante, B., Maes, D., Sibila, M., Betlach, A.M., Sponheim, A. *et al.* (2022) Improving *Mycoplasma hyopneumoniae* diagnostic capabilities by harnessing the infection dynamics. *Veterinary Journal* 288, 105877.

Garrido, F., Leon, L., Ladero, J.L., Cuellar, L. and DIAZ, M.A. (1987) Contagious agalactia in Spain. In: Jones, G.E. (ed.) *Agriculture: Contagious Agalactia and Other Mycoplasmal Diseases of Small Ruminants*. Commission of the European Communities, Brussels, pp. 1–5.

Garza-Moreno, L., Vilalta, C. and Pieters, M. (2022) Environmental detection of *Mycoplasma hyopneumoniae* in breed-to-wean farms. *Research in Veterinary Science* 145, 188–192.

Gates, M., Woolhouse, M., Gunn, G. and Humphry, R. (2013) Relative associations of cattle movements, local spread, and biosecurity with bovine viral diarrhoea virus (BVDV) seropositivity in beef and dairy herds. *Preventive Veterinary Medicine* 112(3–4), 285–295.

Gautier-Bouchardon, A.V., Ferré, S., Le Grand, D., Paoli, A., Gay, E. *et al.* (2014) Overall decrease in the susceptibility of *Mycoplasma bovis* to antimicrobials over the past 30 years in France. *PLoS One* 9(2), e87672.

Gay, E. and Barnouin, J. (2009) A nation-wide epidemiological study of acute bovine respiratory disease in France. *Preventive Veterinary Medicine* 89(3–4), 265–271.

Gelgie, A.E., Korsa, M.G. and Kerro Dego, O.K. (2022) *Mycoplasma bovis* mastitis. *Current Research in Microbial Sciences* 3, 100123.

George, D.R., Finn, R.D., Graham, K.M., Mul, M.F., Maurer, V. *et al.* (2015) Should the poultry red mite *Dermanyssus gallinae* be of wider concern for veterinary and medical science. *Parasites & Vectors* 8(1), 178.

Gerhold, R.W., Fuller, A.L., Beckstead, R.B. and McDougald, L.R. (2010) Low-dose immunization of northern bobwhites (*Colinus virginianus*) with *Eimeria lettyae* provides protection against a high-dose challenge. *Avian Diseases* 54(4), 1220–1223.

Giadinis, N., Arsenos, G., Tsakos, P., Psychas, V., Dovas, C. *et al.* (2012) "Milk-drop syndrome of ewes": Investigation of the causes in dairy sheep in Greece. *Small Ruminant Research* 106(1), 33–35.

Gil, A.K.R., Aguirre-Salado, C.A., Fonseca-Rodríguez, O., Ballesteros-Rodea, G., Socorro Hernández-Arteaga, L.E. *et al.* (2022) What do we have learned about mastitis spatial analysis during the last 30 years. *Large Animal Review* 28(3), 153–160.

Gille, L., Callens, J., Supré, K., Boyen, F., Haesebrouck, F. *et al.* (2018) Use of a breeding bull and absence of a calving pen as risk factors for the presence of *Mycoplasma bovis* in dairy herds. *Journal of Dairy Science* 101(9), 8284–8290.

Go, N., Belloc, C., Bidot, C. and Touzeau, S. (2019) Why, when and how should exposure be considered at the within-host scale? A modelling contribution to PRRSv infection. *Mathematical Medicine and Biology* 36(2), 179–206.

Gomes, F. and Henriques, M. (2016) Control of bovine mastitis: Old and recent therapeutic approaches. *Current Microbiology* 72(4), 377–382.

Gonçalves, J., Kamphuis, C., Martins, C., Barreiro, J., Tomazi, T. *et al.* (2018) Bovine subclinical mastitis reduces milk yield and economic return. *Livestock Science* 210, 25–32.

Goodwin, R. (1985) Apparent reinfection of enzootic-pneumonia-free pig herds: Search for possible causes. *Veterinary Record* 116(26), 690–694.

Gorden, P.J. and Plummer, P. (2010) Control, management, and prevention of bovine respiratory disease in dairy calves and cows. *Veterinary Clinics: Food Animal Practice* 26(2), 243–259.

Gottschalk, M. and Broes, A. (2019) Actinobacillosis. In: Zimmerman, J., Karriker, K. and Ramirez, A. (eds) *Diseases of Swine*, 11th edn. Blackwell Publishing, Ames, pp. 749–766.

Gourlay, R.N., Thomas, L.H. and Howard, C.J. (1976) Pneumonia and arthritis in gnotobiotic calves following inoculation with *Mycoplasma agalactiae* subsp *bovis*. *Veterinary Record* 98(25), 506–507.

Graat, E., Jong, M., Frankena, K. and Franken, P. (2001) Modelling the effect of surveillance programmes on spread of bovine herpesvirus 1 between certified cattle herds. *Veterinary Microbiology* 79(3), 193–208.

Graham, D.A. (2013) Bovine herpes virus-1 (bohv-1) in cattle – a review with emphasis on reproductive impacts and the emergence of infection in Ireland and the United Kingdom. *Irish Veterinary Journal* 66(1), 1–12.

Grasland, B., Blanchard, P., Keranflec'h, A., Oger, A., Rose, N. *et al.* (2009) Transmission of porcine circovirus of type 2 (PCV-2) with semen. *Journées de La Recherche Porcine En France* 41, 193–194.

Grau-Roma, L., Hjulsager, C.K., Sibila, M., Kristensen, C.S., López-Soria, S. *et al.* (2009) Infection, excretion and seroconversion dynamics of porcine circovirus type 2 (PCV2) in pigs from post-weaning multisystemic wasting syndrome (PMWS) affected farms in Spain and Denmark. *Veterinary Microbiology* 135(3–4), 272–282.

Greenacre, C.B. and Morishita, T.Y. (2021) *Backyard Poultry Medicine and Surgery: A Guide for Veterinary Practitioners*. John Wiley & Sons.

Greenhalgh, D. and Griffiths, M. (2009a) Backward bifurcation, equilibrium and stability phenomena in a three-stage extended BRSV epidemic model. *Journal of Mathematical Biology* 59(1), 1–36.

Greenhalgh, D. and Griffiths, M. (2009b) Dynamic phenomena arising from an extended core group model. *Mathematical Biosciences* 221(2), 136–149.

Grenfell, B., Smith, G. and Anderson, R. (1987) A mathematical model of the population biology of *Ostertagia ostertagi* in calves and yearlings. *Parasitology* 95 (Pt 2)(2), 389–406.

Grenfell, B.T., Wilson, K., Isham, V., Boyd, H. and Dietz, K. (1995) Modelling patterns of parasite aggregation in natural populations: Trichostrongylid nematode-ruminant interactions as a case study. *Parasitology* 111(S1), 135–151.

Griffiths, M. and Greenhalgh, D. (2011) The probability of extinction in a bovine respiratory syncytial virus epidemic model. *Mathematical Biosciences* 231(2), 144–158.

Groenendaal, H., Zagmutt, F.J., Patton, E.A. and Wells, S.J. (2015) Cost–benefit analysis of vaccination against *Mycobacterium avium* ssp. *paratuberculosis* in dairy cattle, given its cross-reactivity with tuberculosis tests. *Journal of Dairy Science* 98(9), 6070–6084.

Grooms, D.L., Brock, K.V., Bolin, S.R., Grotelueschen, D.M. and Cortese, V.S. (2014) Effect of constant exposure to cattle persistently infected with bovine viral diarrhea virus on morbidity and mortality rates and performance of feedlot cattle. *Journal of the American Veterinary Medical Association* 244(2), 212–224.

Guelbenzu-Gonzalo, M.P., Lozano, J.-M., O'Sullivan, P., Lane, E.A. and Graham, D.A. (2021) A herd investigation tool in support of the Irish bovine viral diarrhoea eradication programme. *Frontiers in Veterinary Science* 8, 694774.

Gunn, G., Heffernan, C., Hall, M., McLeod, A. and Hovi, M. (2008) Measuring and comparing constraints to improved biosecurity amongst GB farmers, veterinarians and the auxiliary industries. *Preventive Veterinary Medicine* 84(3–4), 310–323.

Gutiérrez, C.B., Rodriguez Barbosa, J., Tascón, R., Costa, L., Riera, P. *et al.* (1995) Serological characterisation and antimicrobial susceptibility of *Actinobacillus pleuropneumoniae* strains isolated from pigs in Spain. *Veterinary Record* 137(3), 62–64.

Györke, A., Pop, L. and Cozma, V. (2013) Prevalence and distribution of *Eimeria* species in broiler chicken farms of different capacities. *Parasite* 20, 50.

Haapala, V., Pohjanvirta, T., Vähänikkilä, N., Halkilahti, J., Simonen, H. *et al.* (2018) Semen as a source of *Mycoplasma bovis* mastitis in dairy herds. *Veterinary Microbiology* 216, 60–66.

Haapala, V., Vähänikkilä, N., Kulkas, L., Tuunainen, E., Pohjanvirta, T. *et al.* (2021) *Mycoplasma bovis* infection in dairy herds – risk factors and effect of control measures. *Journal of Dairy Science* 104(2), 2254–2265.

Hale, H., Helmboldt, C., Plastridge, W. and Stula, E. (1962) Bovine mastitis caused by a *Mycoplasma* species. *Cornell Veterinarian* 52, 582–591.

Han, J., Weir, A., Weston, J., Heuer, C. and Gates, M. (2018) Elimination of bovine viral diarrhoea virus in New Zealand: A review of research progress and future directions. *New Zealand Veterinary Journal* 66(6), 273–280.

Han, J.-H., Weston, J.F., Heuer, C. and Gates, M.C. (2020) Modelling the economics of bovine viral diarrhoea virus control in pastoral dairy and beef cattle herds. *Preventive Veterinary Medicine* 182, 105092.

Hananeh, W.M., Al Momani, W., Ababneh, M.M. and Abutarbush, S.M. (2018) *Mycoplasma bovis* arthritis and pneumonia in calves in Jordan: An emerging disease. *Veterinary World* 11(12), 1663–1668.

Harms, P., Sorden, S., Halbur, P., Bolin, S., Lager, K. *et al.* (2001) Experimental reproduction of severe disease in CD/CD pigs concurrently infected with type 2 porcine circovirus and porcine reproductive and respiratory syndrome virus. *Veterinary Pathology* 38(5), 528–539.

Havas, K.A., Makau, D.N., Shapovalov, S., Tolkova, E., VanderWaal, K. *et al.* (2022) A molecular and epidemiological description of a severe porcine reproductive and respiratory syndrome outbreak in a commercial swine production system in Russia. *Viruses* 14(2), 375.

Hawkins, J.A. (1993) Economic benefits of parasite control in cattle. *Veterinary Parasitology* 46(1–4), 159–173.
Hermans, P.G. and Morgan, K.L. (2007) Prevalence and associated risk factors of necrotic enteritis on broiler farms in the United Kingdom; a cross-sectional survey. *Avian Pathology* 36(1), 43–51.
Herrmann-Hoesing, L.M., Broughton-Neiswanger, L.E., Gouine, K.C., White, S.N., Mousel, M.R. *et al.* (2010) Evaluation of a caprine arthritis-encephalitis virus/maedi-visna virus indirect enzyme-linked immunosorbent assay in the serological diagnosis of ovine progressive pneumonia virus in US. sheep. *Clinical and Vaccine Immunology* 17(2), 307–310.
Högberg, N., Hessle, A., Lidfors, L. and Höglund, J. (2023) The effect of weaning age on animal performance in lambs exposed to naturally acquired nematode infections. *Veterinary Parasitology* 316, 109900.
Höglund, J., Dahlström, F., Sollenberg, S. and Hessle, A. (2013) Weight gain-based targeted selective treatments (TST) of gastrointestinal nematodes in first-season grazing cattle. *Veterinary Parasitology* 196(3–4), 358–365.
Holland, C. (2013) *Ascaris:The Neglected Parasite*. Elsevier, Philadelphia.
Holmgren, N., Lundeheim, N. and Wallgren, P. (1999) Infections with *Mycoplasma hyopneumoniae* and *Actinobacillus pleuropneumoniae* in fattening pigs. Influence of piglet production systems and influence on production parameters. *Journal of Veterinary Medicine, Series B* 46(8), 535–544.
Holst, S., Yeske, P. and Pieters, M. (2015) Elimination of *Mycoplasma hyopneumoniae* from breed-to-wean farms: A review of current protocols with emphasis on herd closure and medication. *Journal of Swine Health and Production* 23(6), 321–330.
Hostnik, P., Černe, D., Mrkun, J., Starič, J. and Toplak, I. (2021) Review of infections with bovine herpesvirus 1 in Slovenia. *Frontiers in Veterinary Science* 8, 676549.
Houe, H., Lindberg, A. and Moennig, V. (2006) Moennig V: Test strategies in bovine viral diarrhea virus control and eradication campaigns in Europe. *Journal of Veterinary Diagnostic Investigation* 18(5), 427–436.
Houlihan, M., Veenstra, B., Christian, M., Nicholas, R. and Ayling, R. (2007) Mastitis and arthritis in two dairy herds caused by *Mycoplasma bovis*. *Veterinary Record* 160(4), 126–127.
Howell, A., Baylis, M., Smith, R., Pinchbeck, G. and Williams, D. (2015) Epidemiology and impact of *Fasciola hepatica* exposure in high-yielding dairy herds. *Preventive Veterinary Medicine* 121(1–2), 41–48.
Hurri, E., Ohlson, A., Lundberg, Å., Aspán, A., Pedersen, K. *et al.* (2022) Herd-level prevalence of *Mycoplasma bovis* in Swedish dairy herds determined by antibody ELISA and PCR on bulk tank milk and herd characteristics associated with seropositivity. *Journal of Dairy Science* 105(9), 7764–7772.
Iscaro, C., Cambiotti, V., Petrini, S. and Feliziani, F. (2021) Control programs for infectious bovine rhinotracheitis (IBR) in European countries: An overview. *Animal Health Research Reviews* 22(2), 136–146.
Islam, M.A., Neuhoff, C., Aqter Rony, S., Große-Brinkhaus, C., Uddin, M.J. *et al.* (2019) PBMCs transcriptome profiles identified breed-specific transcriptome signatures for PRRSV vaccination in German Landrace and Pietrain pigs. *PLoS One* 14(9), e0222513.
Islam, Z.U., Bishop, S.C., Savill, N.J., Rowland, R.R., Lunney, J.K. *et al.* (2013) Quantitative analysis of porcine reproductive and respiratory syndrome (PRRS) viremia profiles from experimental infection: A statistical modelling approach. *PLoS One* 8(12), e83567.
Jara, M., Rasmussen, D.A., Corzo, C.A. and Machado, G. (2021) Porcine reproductive and respiratory syndrome virus dissemination across pig production systems in the United States. *Transboundary and Emerging Diseases* 68(2), 667–683.
Jaros, P., McIntyre, L., Morris, R., Johnstone, A., Garkavenko, O. *et al.* (2006) Experimental evidence that an agent other than PCV2 is a necessary cause of PMWS. In: *Proceedings of the 19th International Pig Veterinary Society Congress Proceedings (IPVS'06).*
Jaÿ, M. and Tardy, F. (2019) Contagious agalactia in sheep and goats: Current perspectives. *Veterinary Medicine* 10, 229–247.
Jenkins, M.C., Dubey, J., Miska, K. and Fetterer, R. (2017) Differences in fecundity of *Eimeria maxima* strains exhibiting different levels of pathogenicity in its avian host. *Veterinary Parasitology* 236, 1–6.
Jobert, J.-L., Savoye, C., Cariolet, R., Kobisch, M. and Madec, F. (2000) Experimental aerosol transmission of *Actinobacillus pleuropneumoniae* to pigs. *Canadian Journal of Veterinary Research* 64(1), 21–26.
Jolly, S., Detilleux, J. and Desmecht, D. (2004) Extensive mast cell degranulation in bovine respiratory syncytial virus-associated paroxystic respiratory distress syndrome. *Veterinary Immunology and Immunopathology* 97(3–4), 125–136.

Jones, R.A., Brophy, P.M., Mitchell, E.S. and Williams, H.W. (2017) Rumen fluke (*Calicophoron daubneyi*) on Welsh farms: Prevalence, risk factors and observations on co-infection with *Fasciola hepatica*. *Parasitology* 144(2), 237–247.

Jordan, A., Sadler, R.J., Sawford, K., Andel, M., Ward, M. *et al.* (2021) *Mycoplasma bovis* outbreak in New Zealand cattle: An assessment of transmission trends using surveillance data. *Transboundary and Emerging Diseases* 68(6), 3381–3395.

Jordan, A.G., Citer, L.R., McAloon, C.G., Graham, D., Sergeant, E. *et al.* (2020) Johne's disease in Irish dairy herds: Considerations for an effective national control programme. *Irish Veterinary Journal* 73, 1–14.

Justice-Allen, A., Trujillo, J., Corbett, R., Harding, R., Goodell, G. *et al.* (2010) Survival and replication of mycoplasma species in recycled bedding sand and association with mastitis on dairy farms in Utah. *Journal of Dairy Science* 93(1), 192–202.

Kaba, J., Czopowicz, M., Kuźmak, J., Olech, M., Witkowski, L. *et al.* (2023) A large-scale study on the seroprevalence of small ruminant lentiviral infection in the Polish goat population. *Preventive Veterinary Medicine* 213, 105885.

Kanankege, K.S.T., Machado, G., Zhang, L., Dokkebakken, B., Schumann, V. *et al.* (2019) Use of A voluntary testing program to study the spatial epidemiology of Johne's disease affecting dairy herds in Minnesota: A cross sectional study. *BMC Veterinary Research* 15(1), 11.

Kanda, T., Tanaka, S., Suwanruengsri, M., Sukmawinata, E., Uemura, R. *et al.* (2019) Bovine endocarditis associated with *Mycoplasma bovis*. *Journal of Comparative Pathology* 171, 53–58.

Keefe, G. (2012) Update on control of *Staphylococcus aureus* and *Streptococcus agalactiae* for management of mastitis. *Veterinary Clinics: Food Animal Practice* 28(2), 203–216.

Kenyon, F. and Jackson, F. (2012) Targeted flock/herd and individual ruminant treatment approaches. *Veterinary Parasitology* 186(1–2), 10–17.

Kenyon, F., Sargison, N., Skuce, P. and Jackson, F. (2009) Sheep helminth parasitic disease in South Eastern Scotland arising as a possible consequence of climate change. *Veterinary Parasitology* 163(4), 293–297.

Khan, M.N., Sajid, M.S., Rizwan, H.M., Qudoos, A., Abbas, R.Z. *et al.* (2017) Comparative efficacy of six anthelmintic treatments against natural infection of *Fasciola* species in sheep. *Pakistan Veterinary Journal* 37(1), 65–68.

Kim, J., Chung, H.-K. and Chae, C. (2003) Association of porcine circovirus 2 with porcine respiratory disease complex. *Veterinary Journal* 166(3), 251–256.

Kim, J., Ha, Y., Jung, K., Choi, C. and Chae, C. (2004) Enteritis associated with porcine circovirus 2 in pigs. *Canadian Journal of Veterinary Research* 68(3), 218–221.

Kinde, H., DaMassa, A.J., Wakenell, P.S. and Petty, R. (1994) Mycoplasma infection in a commercial goat dairy caused by *Mycoplasma agalactiae* and *Mycoplasma mycoides* subsp. *mycoides* (Caprine biotype). *Journal of Veterinary Diagnostic Investigation* 6(4), 423–427.

Kirchhoff, J., Uhlenbruck, S., Goris, K., Keil, G.M. and Herrler, G. (2014) Three viruses of the bovine respiratory disease complex apply different strategies to initiate infection. *Veterinary Research* 45(1), 1–12.

Kirkeby, C., Græsbøll, K. and Halasa, T. (2018) Evaluating the impact of transmission mode, calibration level and farmer compliance in simulation models of paratuberculosis in dairy herds. *Scientific Reports* 8(1), 1–12.

Kirkpatrick, B.W. (2010) Genetics of host susceptibility to paratuberculosis. *Paratuberculosis: Organism, Disease, Control* 50–59.

Klein-Jöbstl, D., Arnholdt, T., Sturmlechner, F., Iwersen, M. and Drillich, M. (2015) Results of an online questionnaire to survey calf management practices on dairy cattle breeding farms in Austria and to estimate differences in disease incidences depending on farm structure and management practices. *Acta Veterinaria Scandinavica* 57(1), 44.

Klem, T.B., Gulliksen, S., Lie, K.-I., Løken, T., Østerås, O. *et al.* (2013) Bovine respiratory syncytial virus: Infection dynamics within and between herds. *Veterinary Record* 173(19), 476–476.

Klem, T.B., Sjurseth, S.K., Sviland, S., Gjerset, B., Myrmel, M. *et al.* (2019) Bovine respiratory syncytial virus in experimentally exposed and rechallenged calves: Viral shedding related to clinical signs and the potential for transmission. *BMC Veterinary Research* 15(1), 11.

Klinkenberg, D., Tobias, T., Bouma, A., Leengoed, L., Stegeman, J. *et al.* (2014) Simulation study of the mechanisms underlying outbreaks of clinical disease caused by *Actinobacillus pleuropneumoniae* in finishing pigs. *Veterinary Journal* 202(1), 99–105.

Knubben-Schweizer, G., Rüegg, S., Torgerson, P.R., Rapsch, C., Grimm, F. *et al.* (2010) Control of bovine fasciolosis in dairy cattle in Switzerland with emphasis on pasture management. *Veterinary Journal* 186(2), 188–191.

Konishi, M., Hayama, Y., Shirafuji, H., Kameyama, K.-I., Murakami, K. *et al.* (2016) Serological survey of caprine arthritis-encephalitis virus infection in Japan. *Journal of Veterinary Medical Science* 78(3), 447–450.

Kristensen, C.S., Baadsgaard, N.P. and Toft, N. (2011) A meta-analysis comparing the effect of PCV2 vaccines on average daily weight gain and mortality rate in pigs from weaning to slaughter. *Preventive Veterinary Medicine* 98(4), 250–258.

Kristensen, C.S., Baekbo, P., Bille-Hansen, V., Bøtner, A., Vigre, H. *et al.* (2009) Induction of porcine post-weaning multisystemic wasting syndrome (PMWS) in pigs from PMWS unaffected herds following mingling with pigs from PMWS-affected herds. *Veterinary Microbiology* 138(3–4), 244–250.

Kruse, A.B., Nielsen, L.R. and Alban, L. (2020) Herd typologies based on multivariate analysis of biosecurity, productivity, antimicrobial and vaccine use data from Danish sow herds. *Preventive Veterinary Medicine* 181, 104487.

Kuerpick, B., Schnieder, T. and Strube, C. (2012) Seasonal pattern of *Fasciola hepatica* antibodies in dairy herds in Northern Germany. *Parasitology Research* 111(3), 1085–1092.

Kumar, A., Rahal, A., Chakraborty, S., Verma, A.K. and Dhama, K. (2014) *Mycoplasma agalactiae*, an etiological agent of contagious agalactia in small ruminants: A review. *Veterinary Medicine International* 2014, 286752.

Kupcinskas, T., Stadaliene, I., Paulauskas, A., Trusevicius, P., Petkevicius, S. *et al.* (2017) A comparison of two different anthelmintic treatment regimens against natural gastrointestinal nematode infections on two Lithuanian sheep farms. *Acta Veterinaria Scandinavica* 59(1), 7.

Kuster, K., Cousin, M.-E., Jemmi, T., Schüpbach-Regula, G. and Magouras, I. (2015) Expert opinion on the perceived effectiveness and importance of on-farm biosecurity measures for cattle and swine farms in Switzerland. *PLoS One* 10(12), e0144533.

Labarque, G. and Trinidade, J. (2012) Prevalence of *Mycoplasma hyopneumoniae* infections at weaning age in European pig herds. In: *4th European Symposium of Porcine Health Management*, Pig333.

Lagan Tregaskis, P., Staines, A., Gordon, A., Sheridan, P., McMenamy, M. *et al.* (2021) Co-infection status of novel parvovirus's (PPV2 to 4) with porcine circovirus 2 in porcine respiratory disease complex and porcine circovirus-associated disease from 1997 to 2012. *Transboundary and Emerging Diseases* 68(4), 1979–1994.

Lambert, M. (1987) Contagious agalactia of sheep and goats. *Revue Scientifique et Technique (International Office of Epizootics)* 6(3), 681–711.

Larghi, M. (2018) Comparative study in the control of bovine viral diarrhea. *Animal Health Research Reviews* 19(2), 125–133.

Larsen, L.E., Tjørnehøj, K. and Viuff, B. (2000) Extensive sequence divergence among bovine respiratory syncytial viruses isolated during recurrent outbreaks in closed herds. *Journal of Clinical Microbiology* 38(11), 4222–4227.

LeBlanc, P., Baker, J., Gray, P., Robinson, N. and Derksen, F. (1991) Effects of bovine respiratory syncytial virus on airway function in neonatal calves. *American Journal of Veterinary Research* 52(9), 1401–1406.

LeCann, P., Albina, E., Madec, F., Cariolet, R. and Jestin, A. (1997) Piglet wasting disease. *Veterinary Record* 141(25), 660.

Leginagoikoa, I., Juste, R., Barandika, J., Amorena, B., De Andrés, D. *et al.* (2006) Extensive rearing hinders Maedi-Visna virus (MVV) infection in sheep. *Veterinary Research* 37(6), 767–778.

Leitner, G., Krifucks, O., Weisblit, L., Lavi, Y., Bernstein, S. *et al.* (2010) The effect of caprine arthritis encephalitis virus infection on production in goats. *Veterinary Journal* 183(3), 328–331.

Le Roux, Y., Laurent, F. and Moussaoui, F. (2003) Polymorphonuclear proteolytic activity and milk composition change. *Veterinary Research* 34(5), 629–645.

Lindberg, A., Brownlie, J., Gunn, G., Houe, H., Moening, V. *et al.* (2006) The control of bovine viral diarrhoea virus in Europe: Today and in the future. *Plurithematic Issue of the Scientific and Technical Review* 961–979.

López-Lorenzo, G., Díaz-Cao, J.M., Prieto, A., López-Novo, C., López, C.M. *et al.* (2019) Environmental distribution of porcine circovirus type 2 (PCV2) in swine herds with natural infection. *Scientific Reports* 9(1), 14816.

Lopez-Soria, S., Segalés, J., Rose, N., Viñas, M., Blanchard, P. *et al.* (2005) An exploratory study on risk factors for postweaning multisystemic wasting syndrome (PMWS) in Spain. *Preventive Veterinary Medicine* 69(1–2), 97–107.

Loria, G., Puleio, R. and Nicholas, R. (2018) Contagious agalactia: Economic losses and good practice. *Journal of Bacteriology and Mycology* 5(5), 1076.

Lu, Z., Schukken, Y.H., Smith, R.L. and Gröhn, Y.T. (2013) Using vaccination to prevent the invasion of *Mycobacterium avium* subsp. *paratuberculosis* in dairy herds: A stochastic simulation study. *Preventive Veterinary Medicine* 110(3–4), 335–345.

Lu, Z., Mitchell, R.M., Smith, R.L., Kessel, J.S., Chapagain, P.P. *et al.* (2008) The importance of culling in Johne's disease control. *Journal of Theoretical Biology* 254(1), 135–146.

Lunney, J.K., Benfield, D.A. and Rowland, R.R. (2010) Porcine reproductive and respiratory syndrome virus: An update on an emerging and re-emerging viral disease of swine. *Virus Research* 154(1–2), 1–6.

Lurier, T., Delignette-Muller, M.L., Rannou, B., Strube, C., Arcangioli, M.-A. *et al.* (2018) Diagnosis of bovine dictyocaulosis by bronchoalveolar lavage technique: A comparative study using a Bayesian approach. *Preventive Veterinary Medicine* 154, 124–131.

Luzzago, C., Bronzo, V., Salvetti, S., Frigerio, M. and Ferrari, N. (2010) Bovine respiratory syncytial virus seroprevalence and risk factors in endemic dairy cattle herds. *Veterinary Research Communications* 34(1), 19–24.

Madec, F. and Josse, J. (1984) The risk factors of respiratory diseases on fatteners in intensive breeding-finishing [pig] units. In: *Proceedings of the 8th International Pig Veterinary Society Congress*, Gent, August 27–31.

Madec, F., Rose, N., Grasland, B., Cariolet, R. and Jestin, A. (2008) Post-weaning multisystemic wasting syndrome and other PCV2-related problems in pigs: A 12-year experience. *Transboundary and Emerging Diseases* 55(7), 273–283.

Madson, D.M., Ramamoorthy, S., Kuster, C., Pal, N., Meng, X.J. *et al.* (2009) Infectivity of porcine circovirus type 2 DNA in semen from experimentally-infected boars. *Veterinary Research* 40(1), 10.

Maes, D., Chiers, K., Haesebrouck, F., Laevens, H., Verdonck, M. *et al.* (2001) Herd factors associated with the seroprevalences of *Actinobacillus pleuropneumoniae* serovars 2, 3 and 9 in slaughter pigs from farrow-to-finish pig herds. *Veterinary Research* 32(5), 409–419.

Maes, D., Segales, J., Meyns, T., Sibila, M., Pieters, M. *et al.* (2008) Control of *Mycoplasma hyopneumoniae* infections in pigs. *Veterinary Microbiology* 126(4), 297–309.

Makau, D.N., Alkhamis, M.A., Paploski, I.A.D., Corzo, C.A., Lycett, S. *et al.* (2021) Integrating animal movements with phylogeography to model the spread of PRRSV in the USA. *Virus Evolution* 7(2), veab060.

Makoschey, B. and Berge, A.C. (2021) Review on bovine respiratory syncytial virus and bovine parainfluenza – usual suspects in bovine respiratory disease – a narrative review. *BMC Veterinary Research* 17(1), 1–18.

Mars, M., Bruschke, C. and Oirschot, J. (1999) Airborne transmission of BHV1, BRSV, and BVDV among cattle is possible under experimental conditions. *Veterinary Microbiology* 66(3), 197–207.

Martínez-Valladares, M., Martínez-Pérez, J.M., Robles-Pérez, D., Cordero-Pérez, C., Famularo, M.R. *et al.* (2013) The present status of anthelmintic resistance in gastrointestinal nematode infections of sheep in the Northwest of Spain by in vivo and in vitro techniques. *Veterinary Parasitology* 191(1–2), 177–181.

Matos, R.A., Santos, S.B., Alves, R.V., Silva, E.J., Marinho, M.L. *et al.* (2019) Occurrence and risk factors associated with *Mycoplasma agalactiae* infection in dairy goat herds of Paraíba state, Brazil. *Pesquisa Veterinária Brasileira* 39(2), 93–98.

Maunsell, F.P., Woolums, A.R., Francoz, D., Rosenbusch, R.F., Step, D.L. *et al.* (2011) *Mycoplasma bovis* infections in cattle. *Journal of Veterinary Internal Medicine* 25(4), 772–783.

May, K., Brügemann, K., König, S. and Strube, C. (2018) The effect of patent *Dictyocaulus viviparus* (re) infections on individual milk yield and milk quality in pastured dairy cows and correlation with clinical signs. *Parasites & Vectors* 11(1), 24.

Mbindyo, C.M., Gitao, G.C. and Mulei, C.M. (2020) Prevalence, etiology, and risk factors of mastitis in dairy cattle in Embu and Kajiado counties, Kenya. *Veterinary Medicine International* 8831172.

McAloon, C.G., Roche, S., Ritter, C., Barkema, H.W., Whyte, P. *et al.* (2019) A review of paratuberculosis in dairy herds – Part 2: On-farm control. *Veterinary Journal* 246, 54–58.

McCann, C., Baylis, M. and Williams, D.J. (2010) The development of linear regression models using environmental variables to explain the spatial distribution of *Fasciola hepatica* infection in dairy herds in England and Wales. *International Journal for Parasitology* 40(9), 1021–1028.

McCarthy, M.-C., O'Grady, L., McAloon, C.G. and Mee, J.F. (2021) Longitudinal prevalence of antibodies to endemic pathogens in bulk tank milk samples from dairy herds engaged or not in contract heifer rearing. *Frontiers in Veterinary Science* 8, 785128.

Mesa-Pineda, C., Navarro-Ruíz, J.L., López-Osorio, S., Chaparro-Gutiérrez, J.J. and Gómez-Osorio, L.M. (2021) Chicken coccidiosis: From the parasite lifecycle to control of the disease. *Frontiers in Veterinary Science* 8, 787653.

Meyns, T., Maes, D., Dewulf, J., Vicca, J., Haesebrouck, F. *et al*. (2004) Quantification of the spread of *Mycoplasma hyopneumoniae* in nursery pigs using transmission experiments. *Preventive Veterinary Medicine* 66(1), 265–275.

Mezo, M., González-Warleta, M., Castro-Hermida, J.A. and Ubeira, F.M. (2008) Evaluation of the flukicide treatment policy for dairy cattle in Galicia (NW spain). *Veterinary Parasitology* 157(3–4), 235–243.

Michiels, A., Piepers, S., Ulens, T., Van Ransbeeck, N., Del Pozo Sacristán, R. *et al*. (2015) Impact of particulate matter and ammonia on average daily weight gain, mortality and lung lesions in pigs. *Preventive Veterinary Medicine* 121(1–2), 99–107.

Michiels, A., Vranckx, K., Piepers, S., Del Pozo Sacristán, R., Arsenakis, I. *et al*. (2017) Impact of diversity of *Mycoplasma hyopneumoniae* strains on lung lesions in slaughter pigs. *Veterinary Research* 48(1), 2.

Middleton, J., Fox, L., Gay, J., Tyler, J. and Besser, T. (2002) Use of pulsed-field gel electrophoresis for detecting differences in *Staphylococcus aureus* strain populations between dairy herds with different cattle importation practices. *Epidemiology and Infection* 129(2), 387–395.

Minardi da Cruz, J.C., Singh, D.K., Lamara, A. and Chebloune, Y. (2013) Small ruminant lentiviruses (SRLVs) break the species barrier to acquire new host range. *Viruses* 5(7), 1867–1884.

Minguijón, E., Reina, R., Pérez, M., Polledo, L., Villoria, M. *et al*. (2015) Small ruminant lentivirus infections and diseases. *Veterinary Microbiology* 181(1–2), 75–89.

Moennig, V. and Becher, P. (2018) Control of bovine viral diarrhea. *Pathogens* 7(1), 29.

Mohd Nor, N., Steeneveld, W., Mourits, M.C.M. and Hogeveen, H. (2012) Estimating the costs of rearing young dairy cattle in the Netherlands using a simulation model that accounts for uncertainty related to diseases. *Preventive Veterinary Medicine* 106(3–4), 214–224.

Mortensen, S., Stryhn, H., Søgaard, R., Boklund, A., Stärk, K.D. *et al*. (2002) Risk factors for infection of sow herds with porcine reproductive and respiratory syndrome (PRRS) virus. *Preventive Veterinary Medicine* 53(1–2), 83–101.

Mõtus, K., Rilanto, T., Viidu, D.-A., Orro, T. and Viltrop, A. (2021) Seroprevalence of selected endemic infectious diseases in large-scale Estonian dairy herds and their associations with cow longevity and culling rates. *Preventive Veterinary Medicine* 192, 105389.

Mul, M., Niekerk, T., Chirico, J., Maurer, V., Kilpinen, O. *et al*. (2009) Control methods for *Dermanyssus gallinae* in systems for laying hens: Results of an international seminar. *World Poultry Science Journal* 65(4), 589–600.

Mul, M.F. and Koenraadt, C.J. (2009) Preventing introduction and spread of *Dermanyssus gallinae* in poultry facilities using the HACCP method. *Control of Poultry Mites* 167–181.

Murai, K. and Higuchi, H. (2019) Prevalence and risk factors of *Mycoplasma bovis* infection in dairy farms in northern Japan. *Research in Veterinary Science* 123, 29–31.

Murray, G., Cassidy, J., Clegg, T., Tratalos, J., McClure, J. *et al*. (2016) A retrospective epidemiological analysis of risk factors for a primary necropsy diagnosis of bovine respiratory disease. *Preventive Veterinary Medicine* 132, 49–56.

Naciri, M., Mancassola, R., Fort, G., Danneels, B. and Verhaeghe, J. (2011) Efficacy of amine-based disinfectant KENO™COX on the infectivity of *Cryptosporidium parvum* oocysts. *Veterinary Parasitology* 179(1–3), 43–49.

Nandi, S., Kumar, M., Manohar, M. and Chauhan, R. (2009) Bovine herpes virus infections in cattle. *Animal Health Research Reviews* 10(1), 85–98.

Nansen, P. and Roepstorff, A. (1999) Parasitic helminths of the pig: Factors influencing transmission and infection levels. *International Journal for Parasitology* 29(6), 877–891.

Naranjo-Lucena, A., Munita Corbalán, M.P., Martínez-Ibeas, A.M., McGrath, G., Murray, G. *et al*. (2018) Spatial patterns of *Fasciola hepatica* and *Calicophoron daubneyi* infections in ruminants in Ireland and modelling of *C. daubneyi* infection. *Parasites & Vectors* 11(1), 1–13.

Nathues, H., Woeste, H., Doehring, S., Fahrion, A.S., Doherr, M.G. *et al.* (2013) Herd specific risk factors for *Mycoplasma hyopneumoniae* infections in suckling pigs at the age of weaning. *Acta Veterinaria Scandinavica* 55(1), 1–13.

Nathues, H., Fournie, G., Wieland, B., Pfeiffer, D.U. and Stärk, K.D. (2016) Modelling the within-herd transmission of *Mycoplasma hyopneumoniae* in closed pig herds. *Porcine Health Management* 2(1), 10.

Nathues, H., Alarcon, P., Rushton, J., Jolie, R., Fiebig, K. *et al.* (2017) Cost of porcine reproductive and respiratory syndrome virus at individual farm level – an economic disease model. *Preventive Veterinary Medicine* 142, 16–29.

Nathues, H., Alarcon, P., Rushton, J., Jolie, R., Fiebig, K. *et al.* (2018) Modelling the economic efficiency of using different strategies to control porcine reproductive and respiratory syndrome at herd level. *Preventive Veterinary Medicine* 152, 89–102.

Ndahetuye, J.B., Twambazimana, J., Nyman, A.-K., Karege, C., Tukei, M. *et al.* (2020) A cross sectional study of prevalence and risk factors associated with subclinical mastitis and intramammary infections, in dairy herds linked to milk collection centers in Rwanda. *Preventive Veterinary Medicine* 179, 105007.

Nemes, I., Molnár, T., Abonyi, T., Terjék, Z., Bálint, Á. *et al.* (2019) Eradication of PRRS from backyard swine herds in Hungary between 2012 and 2018. *Acta Veterinaria Hungarica* 67(4), 543–552.

Newcomer, B.W., Chamorro, M.F. and Walz, P.H. (2017) Vaccination of cattle against bovine viral diarrhea virus. *Veterinary Microbiology* 206, 78–83.

Nicholas, R. (2002) Improvements in the diagnosis and control of diseases of small ruminants caused by mycoplasmas. *Small Ruminant Research* 45(2), 145–149.

Nicholas, R., Ayling, R. and Loria, G. (2008) Ovine mycoplasmal infections. *Small Ruminant Research* 76(1–2), 92–98.

Nielsen, E.O., Enøe, C., Jorsal, S., Barfod, K., Svensmark, B. *et al.* (2008) Postweaning multisystemic wasting syndrome in Danish pig herds: Productivity, clinical signs and pathology. *Veterinary Record* 162(16), 505–508.

Noack, S., Chapman, H.D. and Selzer, P.M. (2019) Anticoccidial drugs of the livestock industry. *Parasitology Research* 118(7), 2009–2026.

Nodelijk, G., de Jong, M.C., van Leengoed, L.A., Wensvoort, G., Pol, J.M. *et al.* (2001) A quantitative assessment of the effectiveness of PRRSV vaccination in pigs under experimental conditions. *Vaccine* 19(27), 3636–3644.

Nodelijk, G., Jong, M.C.M., Van Nes, A., Vernooy, J.C.M., Leengoed, L.A.M.G. *et al.* (2000) Introduction, persistence and fade-out of porcine reproductive and respiratory syndrome virus in a Dutch breeding herd: A mathematical analysis. *Epidemiology and Infection* 124(1), 173–182.

Nodelijk, G., Nielen, M., DeJong, M.C.M. and Verheijden, J.H.M. (2003) A review of porcine reproductive and respiratory syndrome virus in Dutch breeding herds: Population dynamics and clinical relevance. *Preventive Veterinary Medicine* 60(1), 37–52.

Noordegraaf, A.V., Buijtels, J., Dijkhuizen, A., Franken, P., Stegeman, J. *et al.* (1998) An epidemiological and economic simulation model to evaluate the spread and control of infectious bovine rhinotracheitis in the Netherlands. *Preventive Veterinary Medicine* 36(3), 219–238.

Nordenfors, H., Höglund, J. and Uggla, A. (1999) Effects of temperature and humidity on oviposition, molting, and longevity of *Dermanyssus gallinae* (Acari: Dermanyssidae). *Journal of Medical Entomology* 36(1), 68–72.

Norström, M., Pfeiffer, D.U. and Jarp, J. (1999) A space-time cluster investigation of an outbreak of acute respiratory disease in Norwegian cattle herds. *Preventive Veterinary Medicine* 47(1–2), 107–119.

Norström, M., Skjerve, E. and Jarp, J. (2000a) Risk factors for epidemic respiratory disease in Norwegian cattle herds. *Preventive Veterinary Medicine* 44(1), 87–96.

Norström, M., Pfeiffer, D.U. and Jarp, J. (2000b) A space–time cluster investigation of an outbreak of acute respiratory disease in Norwegian cattle herds. *Preventive Veterinary Medicine* 47(1–2), 107–119.

Novobilský, A., Novák, J., Björkman, C. and Höglund, J. (2015) Impact of meteorological and environmental factors on the spatial distribution of *Fasciola hepatica* in beef cattle herds in Sweden. *BMC Veterinary Research* 11, 128.

NRC (2003) National Research Council (US) Committee on Diagnosis and Control of Johne's Disease. Diagnosis and Control of Johne's Disease. Washington (DC). National Academies Press (US). PMID: 25032299.

O'Grady, L., O'Neill, R., Collins, D., Clegg, T. and More, S. (2008) Herd and within-herd bohv-1 prevalence among Irish beef herds submitting bulls for entry to a performance testing station. *Irish Veterinary Journal* 61(12), 1–7.
Ohlson, A., Heuer, C., Lockhart, C., Tråvén, M., Emanuelson, U. *et al.* (2010) Risk factors for seropositivity to bovine coronavirus and bovine respiratory syncytial virus in dairy herds. *Veterinary Record* 167(6), 201–206.
Ola-Fadunsin, S.D., Uwabujo, P.I., Sanda, I.M., Ganiyu, I.A., Hussain, K. *et al.* (2019) Gastrointestinal helminths of intensively managed poultry in Kwara central, Kwara state, Nigeria: Its diversity, prevalence, intensity, and risk factors. *Veterinary World* 12(3), 389–396.
Oma, V.S., Klem, T., Tråvén, M., Alenius, S., Gjerset, B. *et al.* (2018) Temporary carriage of bovine coronavirus and bovine respiratory syncytial virus by fomites and human nasal mucosa after exposure to infected calves. *BMC Veterinary Research* 14(1), 8.
Opriessnig, T., Thacker, E., Yu, S., Fenaux, M., Meng, X.-J. *et al.* (2004) Experimental reproduction of postweaning multisystemic wasting syndrome in pigs by dual infection with *Mycoplasma hyopneumoniae* and porcine circovirus type 2. *Veterinary Pathology* 41(6), 624–640.
Otake, S., Dee, S., Corzo, C., Oliveira, S. and Deen, J. (2010) Long-distance airborne transport of infectious PRRSV and *Mycoplasma hyopneumoniae* from a swine population infected with multiple viral variants. *Veterinary Microbiology* 145(3–4), 198–208.
Owen, D.G. (1983) The effect of "Alcide" on 4 strains of rodent coccidial oocysts. *Laboratory Animals* 17(4), 267–269.
Papatsiros, V.G., Papakonstantinou, G., Meletis, E., Tsekouras, N., Maragkakis, G. *et al.* (2022) Occurrence and associated risk factors of porcine reproductive and respiratory syndrome virus and porcine circovirus type 2 infections in Greece. *Viral Immunology* 35(3), 200–211.
Paramasivam, R., Gopal, D.R., Dhandapani, R., Subbarayalu, R., Elangovan, M.P. *et al.* (2023) Is AMR in dairy products a threat to human health? An updated review on the origin, prevention, treatment, and economic impacts of subclinical mastitis. *Infection and Drug Resistance* 155–178.
Pardon, B., Alliët, J., Boone, R., Roelandt, S., Valgaeren, B. *et al.* (2015) Prediction of respiratory disease and diarrhea in veal calves based on immunoglobulin levels and the serostatus for respiratory pathogens measured at arrival. *Preventive Veterinary Medicine* 120(2), 169–176.
Patterson, A.R., Madson, D.M., Halbur, P.G. and Opriessnig, T. (2011) Shedding and infection dynamics of porcine circovirus type 2 (PCV2) after natural exposure. *Veterinary Microbiology* 149(1–2), 225–229.
Pavlicevic, A., Pavlovic, I., Ratajac, R., Popovic, D., Davidovic, B. *et al.* (2019) Poultry welfare in terms of poultry red mite (*Dermanyssus gallinae*) impact and control. *Biotechnology in Animal Husbandry* 35(1), 1–11.
Peek, H.W. and Landman, W.J.M. (2011) Coccidiosis in poultry: Anticoccidial products, vaccines and other prevention strategies. *Veterinary Quarterly* 31(3), 143–161.
Pérez, M., Biescas, E., Andrés, X., Leginagoikoa, I., Salazar, E. *et al.* (2010) Visna/maedi virus serology in sheep: Survey, risk factors and implementation of a successful control programme in Aragón (Spain). *Veterinary Journal* 186(2), 221–225.
Pettersson, E., Sjölund, M., Dórea, F.C., Lind, E.O., Grandi, G. *et al.* (2021) Gastrointestinal parasites in Swedish pigs: Prevalence and associated risk factors for infection in herds where animal welfare standards are improved. *Veterinary Parasitology* 295, 109459.
Picault, S., Ezanno, P., Smith, K., Amrine, D., White, B. *et al.* (2022) Modelling the effects of antimicrobial metaphylaxis and pen size on bovine respiratory disease in high and low risk fattening cattle. *Veterinary Research* 53(1), 77.
Pieters, M.G. and Maes, D. (2019) Mycoplasmosis. In: Zimmerman, J., Karriker, K. and Ramirez, A. (eds) *Diseases of Swine*, 11th edn. Blackwell, Ames, pp. 863–883.
Pieters, M., Pijoan, C., Fano, E. and Dee, S. (2009) An assessment of the duration of *Mycoplasma hyopneumoniae* infection in an experimentally infected population of pigs. *Veterinary Microbiology* 134(3–4), 261–266.
Pieters, M., Cline, G.S., Payne, B.J., Prado, C., Ertl, J.R. *et al.* (2014) Intra-farm risk factors for *Mycoplasma hyopneumoniae* colonization at weaning age. *Veterinary Microbiology* 172(3–4), 575–580.
Pileri, E. and Mateu, E. (2016) Review on the transmission porcine reproductive and respiratory syndrome virus between pigs and farms and impact on vaccination. *Veterinary Research* 47(1), 108.
Pinho, L., Thompson, G., Machado, M. and Carvalheira, J. (2013) Management practices associated with the bulk tank milk prevalence of *Mycoplasma* spp. in dairy herds in Northwestern Portugal. *Preventive Veterinary Medicine* 108(1), 21–27.

Pinior, B., Garcia, S., Minviel, J.J. and Raboisson, D. (2019) Epidemiological factors and mitigation measures influencing production losses in cattle due to bovine viral diarrhoea virus infection: A meta-analysis. *Transboundary and Emerging Diseases* 66(6), 2426–2439.

Pitkin, A., Deen, J. and Dee, S. (2009a) Further assessment of fomites and personnel as vehicles for the mechanical transport and transmission of porcine reproductive and respiratory syndrome virus. *Canadian Journal of Veterinary Research* 73(4), 298–302.

Pitkin, A., Deen, J. and Dee, S. (2009b) Use of a production region model to assess the airborne spread of porcine reproductive and respiratory syndrome virus. *Veterinary Microbiology* 136(1–2), 1–7.

Pittman, J.S., Shepherd, G., Thacker, B.J. and Myers, G.H. (2010) *Trichuris suis* in finishing pigs: Case report and review. *Journal of Swine Health and Production* 18(6), 306–313.

Porrero, M.C., Hasman, H., Vela, A.I., Fernández-Garayzábal, J.F., Domínguez, L. *et al.* (2012) Clonal diversity of *Staphylococcus aureus* originating from the small ruminants goats and sheep. *Veterinary Microbiology* 156(1–2), 157–161.

Pothmann, H., Spergser, J., Elmer, J., Prunner, I., Iwersen, M. *et al.* (2015) Severe *Mycoplasma bovis* outbreak in an Austrian dairy herd. *Journal of Veterinary Diagnostic Investigation* 27(6), 777–783.

Pratelli, A., Cirone, F., Capozza, P., Trotta, A., Corrente, M. *et al.* (2021) Bovine respiratory disease in beef calves supported long transport stress: An epidemiological study and strategies for control and prevention. *Research in Veterinary Science* 135, 450–455.

Pujols, J., Lorca-Oró, C., Díaz, I., Russell, L.E., Campbell, J.M. *et al.* (2011) Commercial spray-dried porcine plasma does not transmit porcine circovirus type 2 in weaned pigs challenged with porcine reproductive and respiratory syndrome virus. *Veterinary Journal* 190(2), e16–e20.

Punyapornwithaya, V., Fox, L.K., Hancock, D.D., Gay, J.M. and Alldredge, J.R. (2012) Time to clearance of *Mycoplasma* mastitis: The effect of management factors including milking time hygiene and preferential culling. *Canadian Veterinary Journal* 53(10), 1119–1122.

Qi, L., Beaunée, G., Arnoux, S., Dutta, B.L., Joly, A. *et al.* (2019) Neighbourhood contacts and trade movements drive the regional spread of bovine viral diarrhoea virus (BVDV). *Veterinary Research* 50(1), 30.

Rabbi, A., Islam, A., Majumder, S., Anisuzzaman, A. and Rahman, M. (2006) Gastrointestinal helminths infection in different types of poultry. *Bangladesh Journal of Veterinary Medicine* 4(1), 13–18.

Rajeev, M., Mutinda, M. and Ezenwa, V.O. (2017) Pathogen exposure in cattle at the livestock–wildlife interface. *EcoHealth* 14(3), 542–551.

Rappe, J.C.F., García-Nicolás, O., Flückiger, F., Thür, B., Hofmann, M.A. *et al.* (2016) Heterogeneous antigenic properties of the porcine reproductive and respiratory syndrome virus nucleocapsid. *Veterinary Research* 47(1), 117.

Rapsch, C., Dahinden, T., Heinzmann, D., Torgerson, P.R., Braun, U. *et al.* (2008) An interactive map to assess the potential spread of *Lymnaea truncatula* and the free-living stages of *Fasciola hepatica* in Switzerland. *Veterinary Parasitology* 154(3–4), 242–249.

Rashid, M., Rashid, M.I., Akbar, H., Ahmad, L., Hassan, M.A. *et al.* (2019) A systematic review on modelling approaches for economic losses studies caused by parasites and their associated diseases in cattle. *Parasitology* 146(2), 129–141.

Rasmussen, P., Barkema, H.W., Mason, S., Beaulieu, E. and Hall, D.C. (2021) Economic losses due to Johne's disease (paratuberculosis) in dairy cattle. *Journal of Dairy Science* 104(3), 3123–3143.

Rathkjen, P.H. and Dall, J. (2017) Control and eradication of porcine reproductive and respiratory syndrome virus type 2 using a modified-live type 2 vaccine in combination with a load, close, homogenise model: An area elimination study. *Acta Veterinaria Scandinavica* 59, 1–12.

Reina, R., Berriatua, E., Luján, L., Juste, R., Sánchez, A. *et al.* (2009) Prevention strategies against small ruminant lentiviruses: An update. *Veterinary Journal* 182(1), 31–37.

Renault, V., Damiaans, B., Sarrazin, S., Humblet, M.-F., Lomba, M. *et al.* (2018) Classification of adult cattle infectious diseases: A first step towards prioritization of biosecurity measures. *Transboundary and Emerging Diseases* 65(6), 1991–2005.

Renken, C., Nathues, C., Swam, H., Fiebig, K., Weiss, C. *et al.* (2021) Application of an economic calculator to determine the cost of porcine reproductive and respiratory syndrome at farm-level in 21 pig herds in Germany. *Porcine Health Management* 7(1), 3.

Richter, V., Lebl, K., Baumgartner, W., Obritzhauser, W., Käsbohrer, A. *et al.* (2017) A systematic worldwide review of the direct monetary losses in cattle due to bovine viral diarrhoea virus infection. *Veterinary Journal* 220, 80–87.

Rinaldi, L., Catelan, D., Musella, V., Cecconi, L., Hertzberg, H. *et al.* (2015) *Haemonchus contortus*: Spatial risk distribution for infection in sheep in Europe. *Geospatial Health* 9(2), 325–331.

Rodrigues da Costa, M., Fitzgerald, R.M., Manzanilla, E.G., O'Shea, H., Moriarty, J. *et al.* (2020) A cross-sectional survey on respiratory disease in a cohort of Irish pig farms. *Irish Veterinary Journal* 73(1), 24.

Roessler, A.S., Oehm, A.W., Knubben-Schweizer, G. and Groll, A. (2022) A machine learning approach for modelling the occurrence of *Galba truncatula* as the major intermediate host for *Fasciola hepatica* in Switzerland. *Preventive Veterinary Medicine* 200, 105569.

Rose, H., Caminade, C., Bolajoko, M.B., Phelan, P., Dijk, J. *et al.* (2016) Climate-driven changes to the spatio-temporal distribution of the parasitic nematode, *Haemonchus contortus*, in sheep in Europe. *Global Change Biology* 22(3), 1271–1285.

Rose, N., Andraud, M., Grasland, B., Durand, B., Cariolet, R. *et al.* (2009) Experimental quantitative evaluation of porcine circovirus type 2 (PCV-2) transmission. *Interest for Modelling. Journées de La Recherche Porcine En France* 41, 191–192.

Rosendal, T., Widgren, S., Ståhl, K. and Frössling, J. (2020) Modelling spread and surveillance of *Mycobacterium avium* subsp. *paratuberculosis* in the Swedish cattle trade network. *Preventive Veterinary Medicine* 183, 105152.

Rose Vineer, H., Steiner, J., Knapp-Lawitzke, F., Bull, K., von Son-de Fernex, E. *et al.* (2016) Implications of between-isolate variation for climate change impact modelling of *Haemonchus contortus* populations. *Veterinary Parasitology* 229, 144–149.

Rose Vineer, H., Morgan, E.R., Hertzberg, H., Bartley, D.J., Bosco, A. *et al.* (2020a) Increasing importance of anthelmintic resistance in European livestock: Creation and meta-analysis of an open database. *Parasite* 27, 69.

Rose Vineer, H., Verschave, S.H., Claerebout, E., Vercruysse, J., Shaw, D.J. *et al.* (2020b) GLOWORM-PARA: A flexible framework to simulate the population dynamics of the parasitic phase of gastrointestinal nematodes infecting grazing livestock. *International Journal for Parasitology* 50(2), 133–144.

Rossiter, C.A. and Burhans, W.S. (1996) Farm-specific approach to paratuberculosis (Johne's disease) control. *Veterinary Clinics of North America. Food Animal Practice* 12(2), 383–415.

Roy, L. and Buronfosse, T. (2011) Using mitochondrial and nuclear sequence data for disentangling population structure in complex pest species: A case study with *Dermanyssus gallinae*. *PLoS One* 6(7), e22305.

Rypuła, K., Płoneczka-Janeczko, K., Kita, J., Kumala, A. and Zmudziński, J.F. (2012) Seroprevalence of BHV-1 (bovine herpesvirus type 1) among non-vaccinated dairy cattle herds with respiratory disorders. *Polish Journal of Veterinary Sciences* 15(3), 561–563.

Saade, G., Deblanc, C., Bougon, J., Marois-Créhan, C., Fablet, C. *et al.* (2020) Coinfections and their molecular consequences in the porcine respiratory tract. *Veterinary Research* 51(1), 1–9.

Salimi-Bejestani, M., McGarry, J., Felstead, S., Ortiz, P., Akca, A. *et al.* (2005) Development of an antibody-detection ELISA for *Fasciola hepatica* and its evaluation against a commercially available test. *Research in Veterinary Science* 78(2), 177–181.

Sanchez, R.E.J., Nauwynck, H.J., McNeilly, F., Allan, G.M. and Pensaert, M.B. (2001) Porcine circovirus 2 infection in swine foetuses inoculated at different stages of gestation. *Veterinary Microbiology* 83(2), 169–176.

Sanhueza, J.M., Stevenson, M.A., Vilalta, C., Kikuti, M. and Corzo, C.A. (2020) Spatial relative risk and factors associated with porcine reproductive and respiratory syndrome outbreaks in United States breeding herds. *Preventive Veterinary Medicine* 183, 105128.

Santman-Berends, I.M.G.A., Mars, M.H., Waldeck, H.W.F., Van Duijn, L., Wever, P. *et al.* (2018) Quantification of the probability of reintroduction of IBR in the Netherlands through cattle imports. *Preventive Veterinary Medicine* 150, 168–175.

Sarmiento-Silva, R.E., Nakamura-Lopez, Y. and Vaughan, G. (2012) Epidemiology, molecular epidemiology and evolution of bovine respiratory syncytial virus. *Viruses* 4(12), 3452–3467.

Sarrazin, S., Dewulf, J., Mathijs, E., Laureyns, J., Mostin, L. *et al.* (2014) Virulence comparison and quantification of horizontal bovine viral diarrhoea virus transmission following experimental infection in calves. *Veterinary Journal* 202(2), 244–249.

Sassu, E.L., Bossé, J.T., Tobias, T.J., Gottschalk, M., Langford, P.R. *et al.* (2018) Update on *Actinobacillus pleuropneumoniae* - knowledge, gaps and challenges. *Transboundary and Emerging Diseases* 65(S1), 72–90.

Savoye, C., Jobert, J., Berthelot-Hérault, F., Keribin, A., Cariolet, R. *et al.* (2000) A PCR assay used to study aerosol transmission of *Actinobacillus pleuropneumoniae* from samples of live pigs under experimental conditions. *Veterinary Microbiology* 73(4), 337–347.

Scarpellini, R., Giacometti, F., Savini, F., Arrigoni, N., Garbarino, C.A. *et al.* (2023) Bovine paratuberculosis: Results of a control plan in 64 dairy farms in a 4-year period. *Preventive Veterinary Medicine* 215, 105923.

Schibrowski, M.L., Gibson, J.S., Hay, K.E., Mahony, T.J. and Barnes, T.S. (2018) *Mycoplasma bovis* and bovine respiratory disease: A risk factor study in Australian feeder cattle. *Preventive Veterinary Medicine* 157, 152–161.

Schnitt, A. and Tenhagen, B.-A. (2020) Risk factors for the occurrence of methicillin-resistant *Staphylococcus aureus* in dairy herds: An update. *Foodborne Pathogens and Disease* 17(10), 585–596.

Schunn, A.-M., Conraths, F.J., Staubach, C., Fröhlich, A., Forbes, A. *et al.* (2013) Lungworm infections in German dairy cattle herds – seroprevalence and GIS-supported risk factor analysis. *PLoS One* 8(9), e74429.

Schweizer, M., Stalder, H., Haslebacher, A., Grisiger, M., Schwermer, H. *et al.* (2021) Eradication of bovine viral diarrhoea (BVD) in cattle in Switzerland: Lessons taught by the complex biology of the virus. *Frontiers in Veterinary Science* 8, 702730.

Sekiguchi, S., Presi, P., Omori, R., Staerk, K., Schuppers, M. *et al.* (2018) Evaluation of bovine viral diarrhoea virus control strategies in dairy herds in Hokkaido, Japan, using stochastic modelling. *Transboundary and Emerging Diseases* 65(1), e135–e144.

Shen, H., Wang, C., Madson, D.M. and Opriessnig, T. (2010) High prevalence of porcine circovirus viremia in newborn piglets in five clinically normal swine breeding herds in North America. *Preventive Veterinary Medicine* 97(3–4), 228–236.

Sherwin, C.M., Nasr, M.A., Gale, E., Petek, M., Stafford, K. *et al.* (2013) Prevalence of nematode infection and faecal egg counts in free-range laying hens: Relations to housing and husbandry. *British Poultry Science* 54(1), 12–23.

Shivaramaiah, C., Barta, J.R., Hernandez-Velasco, X., Téllez, G. and Hargis, B.M. (2014) Coccidiosis: Recent advancements in the immunobiology of *Elmeria* species, preventive measures, and the importance of vaccination as a control tool against these apicomplexan parasites. *Veterinary Medicine* 5, 23–34.

Sibila, M., Nofrarías, M., López-Soria, S., Segalés, J., Valero, O. *et al.* (2007) Chronological study of *Mycoplasma hyopneumoniae* infection, seroconversion and associated lung lesions in vaccinated and non-vaccinated pigs. *Veterinary Microbiology* 122(1–2), 97–107.

Silva, A.P.S.P., Marostica, T.P., McDaniel, A., Arruda, B.L., Alonso, C. *et al.* (2021) Comparison of *Mycoplasma hyopneumoniae* response to infection by route of exposure. *Veterinary Microbiology* 258, 109118.

Silva, G.S., Corbellini, L.G., Linhares, D.L.C., Baker, K.L. and Holtkamp, D.J. (2018) Development and validation of a scoring system to assess the relative vulnerability of swine breeding herds to the introduction of PRRS virus. *Preventive Veterinary Medicine* 160, 116–122.

Silva, G.S., Machado, G., Baker, K.L., Holtkamp, D.J. and Linhares, D.C.L. (2019) Machine-learning algorithms to identify key biosecurity practices and factors associated with breeding herds reporting PRRS outbreak. *Preventive Veterinary Medicine* 171, 104749.

Smith, G. and Grenfell, B. (1994) Modelling of parasite populations: Gastrointestinal nematode models. *Veterinary Parasitology* 54(1–3), 127–143.

Smith, R.L., Schukken, Y.H. and Gröhn, Y.T. (2015) A new compartmental model of *Mycobacterium avium* subsp. *paratuberculosis* infection dynamics in cattle. *Preventive Veterinary Medicine* 122(3), 298–305.

Smith, R.L., Al-Mamun, M.A. and Gröhn, Y.T. (2017) Economic consequences of paratuberculosis control in dairy cattle: A stochastic modeling study. *Preventive Veterinary Medicine* 138, 17–27.

Snowder, G. (2009) Genetics, environment and bovine respiratory disease. *Animal Health Research Reviews* 10(2), 117–119.

Snowder, G.D., Van Vleck, L.D., Cundiff, L.V. and Bennett, G.L. (2006) Bovine respiratory disease in feedlot cattle: Environmental, genetic, and economic factors. *Journal of Animal Science* 84(8), 1999–2008.

Snowder, G.D., Van Vleck, L.D., Cundiff, L.V., Bennett, G.L., Koohmaraie, M. *et al.* (2007) Bovine respiratory disease in feedlot cattle: Phenotypic, environmental, and genetic correlations with growth, carcass, and longissimus muscle palatability traits. *Journal of Animal Science* 85(8), 1885–1892.

Sparagano, O., Pavlićević, A., Murano, T., Camarda, A., Sahibi, H. *et al.* (2009) Prevalence and key figures for the poultry red mite *Dermanyssus gallinae* infections in poultry farm systems. *Experimental & Applied Acarology* 48(1–2), 3–10.

Sparagano, O.A.E., George, D.R., Harrington, D.W.J. and Giangaspero, A. (2014) Significance and control of the poultry red mite, *Dermanyssus gallinae*. *Annual Review of Entomology* 59, 447–466.

Stokstad, M., Klem, T.B., Myrmel, M., Oma, V.S., Toftaker, I. *et al.* (2020) Using biosecurity measures to combat respiratory disease in cattle: The Norwegian control program for bovine respiratory syncytial virus and bovine coronavirus. *Frontiers in Veterinary Science* 7, 167.

Strain, S., Verner, S., Campbell, E., Hodnik, J.J. and Santman-Berends, I.M.G.A. (2021) The Northern Ireland control programmes for infectious cattle diseases not regulated by the EU. *Frontiers in Veterinary Science* 8, 694197.

Streeter, R.N., Hoffsis, G., Bech-Nielsen, S., Shulaw, W. and Rings, D. (1995) Isolation of *Mycobacterium paratuberculosis* from colostrum and milk of subclinically infected cows. *American Journal of Veterinary Research* 56(10), 1322–1324.

Su, N., Wang, Q., Liu, H.-Y., Li, L.-M., Tian, T. *et al.* (2022) Prevalence of bovine viral diarrhea virus in cattle between 2010 and 2021: A global systematic review and meta-analysis. *Frontiers in Veterinary Science* 9, 1086180.

Sweeney, R.W., Whitlock, R.H. and Rosenberger, A.E. (1992) *Mycobacterium paratuberculosis* cultured from milk and supramammary lymph nodes of infected asymptomatic cows. *Journal of Clinical Microbiology* 30(1), 166–171.

Taylor, J.D., Fulton, R.W., Lehenbauer, T.W., Step, D.L. and Confer, A.W. (2010) The epidemiology of bovine respiratory disease: What is the evidence for preventive measures? *Canadian Veterinary Journal* 51(12), 1351–1359.

Thamsborg, S.M., Nejsum, P. and Mejer, H. (2013) Impact of *Ascaris suum* in livestock. In: Holland, C. (ed.) *Ascaris:The Neglected Parasite*. Elsevier, Philadelphia, pp. 363–381.

Timonen, A.A., Katholm, J., Petersen, A., Mõtus, K. and Kalmus, P. (2017) Within-herd prevalence of intramammary infection caused by *Mycoplasma bovis* and associations between cow udder health, milk yield, and composition. *Journal of Dairy Science* 100(8), 6554–6561.

Timonen, A.A., Autio, T., Pohjanvirta, T., Häkkinen, L., Katholm, J. *et al.* (2020) Dynamics of the within-herd prevalence of *Mycoplasma bovis* intramammary infection in endemically infected dairy herds. *Veterinary Microbiology* 242, 108608.

Timsit, E., Arcangioli, M.-A., Bareille, N., Seegers, H. and Assié, S. (2012) Transmission dynamics of *Mycoplasma bovis* in newly received beef bulls at fattening operations. *Journal of Veterinary Diagnostic Investigation* 24(6), 1172–1176.

Todaro, M., Puleio, R., Sabelli, C., Scatassa, M., Console, A. *et al.* (2015) Determination of milk production losses in Valle del Belice sheep following experimental infection of *Mycoplasma agalactiae*. *Small Ruminant Research* 123(1), 167–172.

Toftaker, I., Sanchez, J., Stokstad, M. and Nødtvedt, A. (2016) Bovine respiratory syncytial virus and bovine coronavirus antibodies in bulk tank milk – risk factors and spatial analysis. *Preventive Veterinary Medicine* 133, 73–83.

Tonni, M., Formenti, N., Boniotti, M.B., Guarneri, F., Scali, F. *et al.* (2022) The role of co-infections in *M. hyopneumoniae* outbreaks among heavy fattening pigs: A field study. *Veterinary Research* 53(1), 41.

Townsend, H., Meek, A.H., Lesnick, T.G. and Janzen, E.D. (1989) Factors associated with average daily gain, fever and lameness in beef bulls at the Saskatchewan central feed test station. *Canadian Journal of Veterinary Research* 53(3), 349–354.

Troell, K., Waller, P. and Höglund, J. (2005) The development and overwintering survival of free-living larvae of *Haemonchus contortus* in Sweden. *Journal of Helminthology* 79(4), 373–379.

Urban-Chmiel, R., Wernicki, A., Puchalski, A., Dec, M., Stęgierska, D. *et al.* (2015) Detection of bovine respiratory syncytial virus infections in young dairy and beef cattle in Poland. *Veterinary Quarterly* 35(1), 33–36.

Vähänikkilä, N., Pohjanvirta, T., Haapala, V., Simojoki, H., Soveri, T. *et al.* (2019) Characterisation of the course of *Mycoplasma bovis* infection in naturally infected dairy herds. *Veterinary Microbiology* 231, 107–115.

Valarcher, J.-F., Schelcher, F. and Bourhy, H. (2000) Evolution of bovine respiratory syncytial virus. *Journal of Virology* 74(22), 10714–10728.

Valarcher, J.-F. and Taylor, G. (2007) Bovine respiratory syncytial virus infection. *Veterinary Research* 38(2), 153–180.

van der Fels-Klerx, H., Sørensen, J., Jalvingh, A. and Huirne, R. (2001) An economic model to calculate farm-specific losses due to bovine respiratory disease in dairy heifers. *Preventive Veterinary Medicine* 51(1–2), 75–94.

VanderWaal, K., Paploski, I.A., Makau, D.N., Corzo, C.A. and Paploski, I.A. (2020) Contrasting animal movement and spatial connectivity networks in shaping transmission pathways of a genetically diverse virus. *Preventive Veterinary Medicine* 178, 104977.

Van Emous, R. (2005) Wage war against the red mite!. *Poultry International* 44, 26.

Van Limbergen, T., Sarrazin, S., Chantziaras, I., Dewulf, J., Ducatelle, R. *et al.* (2020) Risk factors for poor health and performance in European broiler production systems. *BMC Veterinary Research* 16(1), 287.

Van Metre, D.C., Tennant, B.C. and Whitlock, R.H. (2007) Infectious diseases of the gastrointestinal tract. In: Divers, T. and Peek, S. (eds) *Rebhun's Diseases of Dairy Cattle*, 2nd edn. Elsevier, Philadelphia, pp. 200–294.

van Roon, A.M., Madouasse, A., Toft, N., Santman-Berends, I.M., Gethmann, J. *et al.* (2022) Output-based assessment of herd-level freedom from infection in endemic situations: Application of a Bayesian hidden Markov model. *Preventive Veterinary Medicine* 204, 105662.

Vasileiou, N., Fthenakis, G.C. and Papadopoulos, E. (2015) Dissemination of parasites by animal movements in small ruminant farms. *Veterinary Parasitology* 213(1–2), 56–60.

Vass-Bognár, B., Bakony, M., Baumgartner, W., Khol, J.L. and Jurkovich, V. (2022) Association between calf rearing technology and farm-level paratuberculosis infection in Hungarian dairy farms. *Preventive Veterinary Medicine* 207, 105719.

Velasova, M., Alarcon, P., Williamson, S. and Wieland, B. (2012) Risk factors for porcine reproductive and respiratory syndrome virus infection and resulting challenges for effective disease surveillance. *BMC Veterinary Research* 8(1), 1–14.

Verdugo, C., Valdes, M.F. and Salgado, M. (2018) Within-herd prevalence and clinical incidence distributions of *Mycobacterium avium* subspecies *paratuberculosis* infection on dairy herds in Chile. *Preventive Veterinary Medicine* 154, 113–118.

Vicca, J., Stakenborg, T., Maes, D., Butaye, P., Peeters, J. *et al.* (2003) Evaluation of virulence of *Mycoplasma hyopneumoniae* field isolates. *Veterinary Microbiology* 97(3–4), 177–190.

Viet, A.-F., Fourichon, C., Jacob, C., Guihenneuc-Jouyaux, C. and Seegers, H. (2006) Approach for qualitative validation using aggregated data for a stochastic simulation model of the spread of the bovine viral-diarrhoea virus in a dairy cattle herd. *Acta Biotheoretica* 54(3), 207–217.

Viet, A.-F., Fourichon, C. and Seegers, H. (2007) Review and critical discussion of assumptions and modelling options to study the spread of the bovine viral diarrhoea virus (BVDV) within a cattle herd. *Epidemiology and Infection* 135(5), 706–721.

Vigre, H., Dohoo, I.R., Stryhn, H. and Busch, M.E. (2004) Intra-unit correlations in seroconversion to *Actinobacillus pleuropneumoniae* and *Mycoplasma hyopneumoniae* at different levels in Danish multi-site pig production facilities. *Preventive Veterinary Medicine* 63(1–2), 9–28.

Vigre, H., Baekbo, P., Jorsal, S.E., Bille-Hansen, V., Hassing, A.-G. *et al.* (2005) Spatial and temporal patterns of pig herds diagnosed with postweaning multisystemic wasting syndrome (PMWS) during the first two years of its occurrence in Denmark. *Veterinary Microbiology* 110(1–2), 17–26.

Villaamil, F.J., Arnaiz, I., Allepuz, A., Molins, M., Lazaro, M. *et al.* (2020) A survey of biosecurity measures and serological status for bovine viral diarrhoea virus and bovine herpesvirus 1 on dairy cattle farms in North-West and North-East Spain. *Veterinary Record Open* 7(1), e000399.

Villoria, M., Leginagoikoa, I., Luján, L., Pérez, M., Salazar, E. *et al.* (2013) Detection of small ruminant lentivirus in environmental samples of air and water. *Small Ruminant Research* 110(2–3), 155–160.

Waap, H., Nunes, T., Mul, M.F., Gomes, J. and Bartley, K. (2019) Survey on the prevalence of *Dermanyssus gallinae* in commercial laying farms in Portugal. *Avian Pathology* 48(suppl 1), S2–S9.

Waldeck, H., Duijn, L., Mar, M.H., Santman-Berends, I.M., Biesheuvel, M.M. *et al.* (2021) Risk factors for introduction of bovine herpesvirus 1 (bohv-1) into cattle herds: A systematic European literature review. *Frontiers in Veterinary Science* 8, 1212.

Wang, T., Vineer, H.R., Morrison, A., van Wyk, J.A., Bolajoko, M.-B. *et al.* (2018) Microclimate has a greater influence than macroclimate on the availability of infective *Haemonchus contortus* larvae on herbage in a warmed temperate environment. *Agriculture, Ecosystems & Environment* 265, 31–36.

Wang, T., Vineer, H.R., Redman, E., Morosetti, A., Chen, R. *et al.* (2022) An improved model for the population dynamics of cattle gastrointestinal nematodes on pasture: Parameterisation and field validation for *Ostertagia ostertagi* and *Cooperia oncophora* in Northern temperate zones. *Veterinary Parasitology* 310, 109777.

Wells, S.J. and Wagner, B.A. (2000) Herd-level risk factors for infection with *Mycobacterium paratuberculosis* in US dairies and association between familiarity of the herd manager with the disease or prior diagnosis of the disease in that herd and use of preventive measures. *Journal of the American Veterinary Medical Association* 216(9), 1450–1457.

Wenke, C., Pospiech, J., Reutter, T., Truyen, U. and Speck, S. (2017) Efficiency of different air filter types for pig facilities at laboratory scale. *PLoS One* 12(10), e0186558.

Wenke, C., Pospiech, J., Reutter, T., Altmann, B., Truyen, U. *et al.* (2018) Impact of different supply air and recirculating air filtration systems on stable climate, animal health, and performance of fattening pigs in a commercial pig farm. *PLoS One* 13(3), e0194641.

Wernike, K., Gethmann, J., Schirrmeier, H., Schröder, R., Conraths, F.J. *et al.* (2017) Six years (2011–2016) of mandatory nationwide bovine viral diarrhea control in Germany – a success story. *Pathogens* 6(4), 50.

Whittington, R., Donat, K., Weber, M.F., Kelton, D., Nielsen, S.S. *et al.* (2019) Control of paratuberculosis: Who, why and how. A review of 48 countries. *BMC Veterinary Research* 15(1), 198.

Whittington, R.J. and Windsor, P.A. (2009) In utero infection of cattle with *Mycobacterium avium* subsp. *paratuberculosis*: A critical review and meta-analysis. *Veterinary Journal* 179(1), 60–69.

Whittington, R.J., Marshall, D.J., Nicholls, P.J., Marsh, I.B. and Reddacliff, L.A. (2004) Survival and dormancy of *Mycobacterium avium* subsp. *paratuberculosis* in the environment. *Applied and Environmental Microbiology* 70(5), 2989–3004.

Williams, R. (1973) Effects of different infection rates on the oocyst production of *Eimeria acervulina* or *Eimeria tenella* in the chicken. *Parasitology* 67(3), 279–288.

Williams, R.B. (2001) Quantification of the crowding effect during infections with the seven *Eimeria* species of the domesticated fowl: Its importance for experimental designs and the production of oocyst stocks. *International Journal for Parasitology* 31(10), 1056–1069.

Wilson, B., Step, D., Maxwell, C., Gifford, C., Richards, C. *et al.* (2017) Effect of bovine respiratory disease during the receiving period on steer finishing performance, efficiency, carcass characteristics, and lung scores. *Professional Animal Scientist* 33(1), 24–36.

Winkler, M., Doster, A. and Jones, C. (2000) Persistence and reactivation of bovine herpesvirus 1 in the tonsils of latently infected calves. *Journal of Virology* 74(11), 5337–5346.

Wolff, C., Emanuelson, U., Ohlson, A., Alenius, S. and Fall, N. (2015) Bovine respiratory syncytial virus and bovine coronavirus in Swedish organic and conventional dairy herds. *Acta Veterinaria Scandinavica* 57(1), 1–7.

Wondlmu, A., Mesfin, E. and Bayu, Y. (2019) Prevalence of poultry coccidiosis and associated risk factors in intensive farming system of Gondar town, Ethiopia. *Veterinary Medicine International* 5748690.

Wongrak, K., Daş, G., Moors, E., Sohnrey, B. and Gauly, M. (2014) Establishment of gastro-intestinal helminth infections in free-range chickens: A longitudinal on farm study. *Berliner Und Munchener Tierarztliche Wochenschrift* 127(7–8), 314–321.

Wongrak, K., Gauly, M. and Daş, G. (2015) Diurnal fluctuations in nematode egg excretion in naturally and in experimentally infected chickens. *Veterinary Parasitology* 208(3–4), 195–203.

Woolley, L.K., Fell, S., Gonsalves, J.R., Walker, M.J., Djordjevic, S.P. *et al.* (2012) Evaluation of clinical, histological and immunological changes and qPCR detection of *Mycoplasma hyopneumoniae* in tissues during the early stages of mycoplasmal pneumonia in pigs after experimental challenge with two field isolates. *Veterinary Microbiology* 161(1–2), 186–195.

Zadoks, R.N. (2002) Molecular and mathematical epidemiology of *Staphylococcus aureus* and *Streptococcus uberis* mastitis in dairy herds. Dissertation, Utrecht University, The Netherlands.

Zadoks, R., Allore, H., Hagenaars, T., Barkema, H. and Schukken, Y. (2002) A mathematical model of *Staphylococcus aureus* control in dairy herds. *Epidemiology and Infection* 129(2), 397–416.

Zajac, A.M. (2006) Gastrointestinal nematodes of small ruminants: Life cycle, anthelmintics, and diagnosis. *Veterinary Clinics of North America. Food Animal Practice* 22(3), 529–541.

Zhou, L., Ge, X. and Yang, H. (2021) Porcine reproductive and respiratory syndrome modified live virus vaccine: A "leaky" vaccine with debatable efficacy and safety. *Vaccines* 9(4), 362.

3 Epidemic Diseases of Livestock

Abstract

In this chapter, epidemic diseases of livestock are examined. For each disease, a brief overview is provided covering clinical signs, economic impact and the significance of the disease to its respective sector. The data relating to the prevalence and presence of ongoing outbreaks of the disease are discussed, with a focus on Europe and near neighbour countries from which there is a risk of spread to Europe. Following this is an overview of risk factors for disease introduction or disease spread. In the case of diseases not currently present in Europe, the risk of introduction to Europe is examined. Transmission routes for the disease are then explored, followed by current information regarding control measures and biosecurity. Finally, an overview of modelling data is provided, with a focus on the availability of existing models of transmission and experimental work delineating parameters of interest for modellers.

Rift Valley Fever

Overview

Rift Valley fever (RVF) is an arthropod-borne viral (arborviral) infection, the aetiological agent of which is the Rift Valley fever virus (RVFV) (Rolin *et al.*, 2013). Following its initial identification in Kenya in 1930, it has been detected across large swathes of the African continent (Bird *et al.*, 2009). Originally endemic to East Africa, it has recently spread and become established across much of Africa, including West Africa, Egypt, Madagascar and Comoros (Rolin *et al.*, 2013). Most recently, RVFV spread beyond Africa to Saudi Arabia and Yemen in 2000 (Balkhy and Memish, 2003; Abdo-Salem *et al.*, 2006).

Rift Valley fever is primarily a disease of domestic animals and wildlife but can also cause sporadic zoonotic outbreaks leading to severe haemorrhagic fever in some cases (Kroeker *et al.*, 2020). Epidemics of RFV typically cause mass abortions and high mortality in ruminants, leading to a significant economic impact (Hartley *et al.*, 2011). Although not currently present in Europe, RVF has been introduced to the French department of Mayotte (Rolin *et al.*, 2013; Métras *et al.*, 2016, 2017) and, thus is technically now endemic to one (extra-European) EU territory. There is also growing concern about the possibility of its introduction due to climate change potentially increasing the range of the arthropod vector species, the close proximity of endemic countries in the Mediterranean basin and ongoing cattle trade with Egypt (Rolin *et al.*, 2013).

Prevalence

Rift Valley fever virus is confirmed as endemic, and causes sporadic outbreaks, in Kenya, Tanzania, Sudan, Somalia, Chad, Zimbabwe, Zambia, South Africa, Eswatini, Namibia, Madagascar, Mauritania, Senegal, the Gambia, Saudi Arabia and Yemen (Rolin *et al.*, 2013), with the most recent outbreaks occurring in Mozambique in 2014 (Fafetine *et al.*, 2016) and Niger (Lagare *et al.*, 2019). RVFV is periodically isolated from, or there exists serological evidence of its presence in, almost all of sub-Saharan Africa except for Ghana, Côte d'Ivoire, Benin, Togo, Guinea-Bissau, Liberia, Sierra Leone, Rwanda, Burundi, Lesotho, Eritrea and Djibouti (Rolin *et al.*, 2013).

Risk factors

Risk factors in the sense discussed elsewhere in this book are not directly applicable to RVF. Instead, this section will focus on risk factors for the introduction of RVFV to Europe.

One of the principal risk factors for the introduction of RVF to Europe is the introduction of viraemic hosts (Hartley *et al.*, 2011; Rolin

Biosecurity for Livestock Diseases in Europe (Alistair Antonopoulos *et al.*)
DOI: 10.1079/ 9781800629837.0003

et al., 2013). Infected ruminants then potentially serve as amplifying hosts, propagating RVFV to domestic vectors and wildlife (Torres-Vélez and Brown, 2004). The introduction of RVFV to Saudi Arabia has been attributed to importation of viraemic livestock from neighbouring countries in the Horn and East Africa (Chevalier *et al.*, 2010). Although there has been some concern for wind-borne dispersal of RVFV vectors, this is considered extremely unlikely except in the case of Morocco to Spain where it remains theoretically feasible but very unlikely, with the risk of this occurring outside an epidemic also being negligible (Rolin *et al.*, 2013). Mechanical transport is considered a much more likely scenario, as the vector species are known to survive being confined within shipping cargo or aircraft holds (Kasari *et al.*, 2008; Hartley *et al.*, 2011).

Transmission

Rift Valley fever is vector borne by mosquitoes of the genera *Aedes*, *Culex* and *Anopheles*, with a total of over 30 species capable of carrying RVFV (Seufi and Galal, 2010; Chevalier *et al.*, 2010). RVF outbreaks typically occur in 7–10 year cycles in endemic regions, potentially due to a natural decrease in seropositive animals from the previous outbreak and/or emergence of particularly conducive climatic conditions for the mosquito vectors (Kroeker *et al.*, 2020). It is thought that *Aedes mcintoshi* mosquitoes initiate the RVFV outbreak (Kroeker *et al.*, 2020) as this species is capable of vertically transmitting and maintaining the virus (Romoser *et al.*, 2011); however, although this is well accepted, it has only been demonstrated in one study (Kroeker *et al.*, 2020). It is also possible that mosquito–animal transmission is continuous and low level, but only becomes observable and manifests as an outbreak when environmental conditions become conducive (Kroeker *et al.*, 2020).

Control measures and biosecurity

Control measures and biosecurity in the sense discussed elsewhere in this book are not directly applicable to RVF; instead, this section will focus on control measures and biosecurity to prevent the introduction of RVF into Europe.

Significant EU legislation and trade restrictions exist designed to prevent the introduction of RVF. All countries exporting livestock to the EU must be officially certified as RVF free, monitored by national veterinary services, and there is a prohibition against trans-shipment through endemic regions (Chevalier *et al.*, 2010). However, cattle trade does occur with Egypt, which has the potential for emerging endemicity of RVFV (Rolin *et al.*, 2013), with the most recent outbreak occurring in 2003 (Kenawy *et al.*, 2018). Another potential route for spread into Europe could be through western Asia and Turkey into southeastern Europe and the Balkans although for the time being, this is not likely as RVFV is only currently endemic in Saudi Arabia and Yemen. It is, however, potentially possible and ecologically feasible for further spread, as similar arboviruses, such as West Nile virus (WNV) have spread to Iran (Rolin *et al.*, 2013).

Another potential route of introduction could be via imported wildlife or zoo animals, although stringent quarantine regulations exist to prevent this from occurring (Chevalier *et al.*, 2010).

Availability of modelling data

A wealth of studies have modelled various aspects of disease transmission, including persistence and control in spatially heterogeneous landscape, wherein R_0 is calculated and reported (Tennant *et al.*, 2021); vertical transmission in vector-borne diseases (Chitnis *et al.*, 2013); changes in land use and their effect on emergence or re-emergence (de la Rocque *et al.*, 2007; Lambin *et al.*, 2010); an open source transmission dynamics model (Sumaye *et al.*, 2019); a dynamic climaed through a sylvatic cycle between Ornithodorus soft ticks and warthogs (Gaudte-driven model (Leedale *et al.*, 2016); the effects of seasonality on transmission (Xiao *et al.*, 2015); heterogeneity of transmission across livestock hosts (Cecilia *et al.*, 2020); local context-specific transmission models focusing on Madagascar (Olive *et al.*, 2016, 2017), Mayotte (Métras *et al.*,

2016, 2017) including the specific role of the movement of livestock (Kim *et al.*, 2021), Tanzania (Sindato *et al.*, 2015, 2016), Kenya (Mosomtai *et al.*, 2016; Ochieng *et al.*, 2016), and the role of livestock movement (Floyd *et al.*, 2019; Mosomtai *et al.*, 2018); Senegal (Lafaye *et al.*, 2013; Durand *et al.*, 2020); the Sahel region (Cecilia *et al.*, 2020) and East Africa (Tran *et al.*, 2016); vector modelling in Saudi Arabia (Naeem *et al.*, 2016) and spatial modelling related to uncertainty management (Clements *et al.*, 2006). Several studies have also modelled the risk of introduction of RVFV to European countries such as The Netherlands (Esser *et al.*, 2020). The transmission potential amongst livestock has also been explored in The Netherlands (Fischer *et al.*, 2013), the UK (Simons *et al.*, 2019) and the US (Golnar *et al.*, 2018). In the study of Fischer *et al.* (2013), the Floquet ratio (R_T) is used in place of R_0 as a more appropriate measure for seasonal systems.

African Swine Fever

Overview

African swine fever (ASF) is an arboviral infection caused by the African swine fever virus (ASFV), originally endemic to Africa where it is maintained through a sylvatic cycle between *Ornithodorus* soft ticks and warthogs (Gaudreault *et al.*, 2020). While the infection is typically asymptomatic in warthogs, ASF in domestic swine is characterized by haemorrhagic fever with the potential for up to 100% mortality in susceptible populations (Blome *et al.*, 2013).

First identified in Kenya in 1921, ASF subsequently spread to Europe. The first recorded outbreak in Western Europe occurred in Portugal in 1957 (Cwynar *et al.*, 2019) and re-emerged in Belgium in 2018 (Linden *et al.*, 2019), whereas in Eastern Europe ASFV was first observed in Georgia in 2007 and spread quickly to neighbouring countries (Cwynar *et al.*, 2019). ASF causes significant economic impact in countries where an outbreak occurs (Brown *et al.*, 2021), with estimates of potential losses of up to €396,468,452 simulated in Denmark (Halasa *et al.*, 2016a, b). Previous analyses estimated the total cost of Spain's 20-year eradication programme, including export and animal losses, at €86,059,643,719 (adjusted for inflation) (Bech-Nielsen *et al.*, 1993; Brown *et al.*, 2021).

Prevalence

As of 2020, there were ongoing outbreaks of ASF in Europe and neighbouring countries: the Russian Federation, Ukraine, Latvia, Poland, Romania, Hungary, Bulgaria, Moldova, Slovakia and Serbia (Gaudreault *et al.*, 2020). Belgium self-declared ASF free in October 2020 (OIE, 2020). ASF was first detected in Germany in September 2020 and is ongoing (Sauter-Louis *et al.*, 2022). The most recent previously resolved outbreaks also occurred in Georgia (2007–2008), Armenia (2010–2011), Azerbaijan (2008), Belarus (2013), Lithuania and Estonia (2014–2019) and Czechia (2018) (Gaudreault *et al.*, 2020). For a thorough analysis of ASFV within the European Union, see the recent EFSA report on epidemiological analysis of ASFV within the EU (EFSA, 2022; 2023; 2024).

Risk factors

Persistence in the tick population is a significant risk factor for spread and re-emergence. ASFV has been isolated from ticks up to 4 years after the culling of all domestic pigs in Madagascar (Ravaomanana *et al.*, 2010), and up to 5 years in ticks collected from farms in Portugal (Boinas *et al.*, 2011). Transmission of ASFV to pigs via infected Iberian ticks has also been demonstrated 588 days post infection of the tick (Endris and Hess, 1992). In the context of the current European outbreak, persistence of ASFV in the wild boar population has been the main driver of ASF spread (Chenais *et al.*, 2019; de la Torre *et al.*, 2022), with farms with low external biosecurity initially being responsible for much of the spread (Chenais *et al.*, 2019). Wild boar movements constitute one of the most significant risk factors for ASFV spread, in addition to the movement of pork and pork products (de la Torre *et al.*, 2022).

Transmission

Ornithodorus ticks serve as both reservoir hosts and biological vectors for transmission of ASFV (Golnar *et al.*, 2019). These ticks have long lifespans and ASFV can multiply to high levels which can be maintained for protracted periods of time with few to no pathological effects in the tick (Kleiboeker *et al.*, 1998; Boinas *et al.*, 2011). However, although these ticks are present in southwestern European countries such as Spain and Portugal, they are unlikely to account for transmission during the ongoing outbreak in Eastern Europe as they are largely absent in Central Europe and the Baltic, and most species in the Caucasus do not infest swine (Frant *et al.*, 2017). ASFV is thought to have been introduced into Georgia via the importation of contaminated pig products from eastern Africa (Rowlands *et al.*, 2008) and its subsequent spread has been attributed to movements of infected wild and domestic pigs (Guinat *et al.*, 2016).

Transmission can occur by contact with infectious animals and fomites, contact with contaminated equipment, vehicles, and clothing (Costard *et al.*, 2013), ingestion of contaminated pig products (Guinat *et al.*, 2016) or contaminated feed (Stoian *et al.*, 2019), with kitchen waste having been identified as a potential source of contamination leading to its feeding to pigs being banned in several countries (Dixon *et al.*, 2020). ASFV is stable under extreme environmental conditions, further facilitating its transmission (Gaudreault *et al.*, 2020), and it can persist in contaminated feed even during trans-Atlantic shipping (Stoian *et al.*, 2019).

Control measures and biosecurity

No vaccines or effective treatments currently exist for ASF (Guinat *et al.*, 2016), so biosecurity remains the only effective measure to prevent an outbreak. Enhanced passive surveillance has also aided in the detection of outbreaks in domestic farms (EFSA, 2022). Given the high mortality rates and rapid spread of ASF within a herd, external biosecurity is of paramount importance (Jurado *et al.*, 2018; Dixon *et al.*, 2020).

The most important preventive measures identified are the identification of animal and farm records; strict enforcement of banning swill feeding; containment of pigs preventing both direct and indirect pig–pig/pig–boar contacts; preventing contact between farm workers and external pigs; removal of carcasses, slaughter residues and food waste; disposal of manure; and not engaging in hunting activity for 48 hours prior to farm visits (Jurado *et al.*, 2018). More general pig production biosecurity measures are also applicable in the case of ASF, such as all-in/all-out systems and proper quarantine management in dedicated units treated as separate from the rest of the farm, showering in and out and the use of proper protective equipment such as boots and gloves (Alarcón *et al.*, 2021).

Finally, the risk of ASFV via animal transport vehicles is also significant and correct washing and disinfection of vehicles, use of optimal disinfectants and ensuring the requisite temperature is reached for safe disinfection are of paramount importance (Gao *et al.*, 2022). Control measures at the wild boar–pig interface are also important, with depopulation, active carcass searching and removal, and the creation of surveillance zones surrounding identified infection zones being recommended measures within the EU (de la Torre *et al.*, 2022).

Availability of modelling data

Guinat *et al.* (2016) carried out experimental infection of pig–pig transmission dynamics for the Georgia 2007/1 strain, calculating a within-pen R_0 of 2.8 and between-pen R_0 of 1.4. A further study estimated R_0 for the 1977 Ukranian epidemic (Korennoy *et al.*, 2017). Barongo *et al.* (2016) calculated R_0 for ASFV transmission between pig herds in Uganda. The importance of persistence and transmission within the wild boar population has been modelled in Europe (Gervasi and Guberti, 2021) and in East Asia (Oberin *et al.*, 2022). Indirect transmission to domestic pig herds in Europe has also been considered (Olesen *et al.*, 2020).

Bergmann *et al.* (2021) examined environmental risk factors related to ASFV in European wild boar. Transmission dynamics within a swine house have been studied by Nielsen *et al.* (2017). Within-herd transmission models have been constructed for Vietnam, where R_0 is calculated for commercial pig farms (Mai *et al.*, 2022a, b) stochastic disease transmission modelling examined time to virus detection (Malladi *et al.*, 2022). Inference of within-herd transmission parameters has been carried out in the context of the ASFV outbreak in the Russian Federation (Guinat *et al.*, 2018). The potential role of haematophagous arthropods in ASFV transmission in France has been studied (Saegerman *et al.*, 2021), in addition to the role of dead carrier pigs (Eblé *et al.*, 2019). Experimental infections have examined indirect and intramuscular injection of ASFV (Yamada *et al.*, 2021) and oral low-dose infection in wild boar (Pietschmann *et al.*, 2015). Experimental modelling has examined ecological niches for ASFV (Hakobyan *et al.*, 2022) and ASFV transmission rate (de Carvalho Ferreira *et al.*, 2013).

African swine fever virus is highly resistant to low temperatures but can be heat inactivated at 56°C for 70 min or >60°C for 20 min. It is inactivated by pH <3.9 or >11.5 in serum-free environments; the presence in serum increases viral resistance to pH up to 13.4 (WOAH Technical Disease Card). Beato *et al.* (2022) and Lorenzi *et al.* (2020) have conducted a thorough review of disinfectants and cleaning procedures against reviewed risk factors (Bergmann *et al.*, 2022). A simulation model was constructed to assess the spread of ASFV including the effects of residues from dead animals (Halasa *et al.*, 2016a) and the epidemiological and economic effects on industrial swine production (Halasa *et al.*, 2016b). A fuzzy model examined epidemic risk in Australia (Liu *et al.*, 2023). Several studies have examined the role of biosecurity in smallholder farms in Africa (Mutua and Dione, 2021; Penrith *et al.*, 2023) and carried out a systematic review and meta-analysis of between-country transmission (Hakizimana *et al.*, 2021). Experimental transmission of ASFV in ticks has also been considered (Endris and Hess, 1992). For a thorough review of the risks of ASFV, classical swine fever and foot and mouth disease (the latter two are both covered below in separate sections), see Brown *et al.* (2020).

Gao *et al.* (2022) conducted a SWOT analysis of risk factors in import/export of pigs within Denmark, considering the importance of disinfection in animal transport, with Neumann *et al.* (2021) carrying out a similar analysis in Australia. An analysis of the risk of ASFV contamination in global feed ingredient supply chains has been carried out (Shurson *et al.*, 2022). Song *et al.* (2022) calculated the basic reproduction number and modelled the stability of ASFV in swill and the environment. Nuanualsuwan *et al.* (2022) examined the persistence of ASFV on porous and non-porous fomites at environmental temperatures. Khanal *et al.* (2022) examined detection of ASFV in experimentally inoculated feed dust. Finally, a mathematical model has been constructed to simulate control measures for ASFV (Barongo *et al.*, 2016).

Peste Des Petits Ruminants

Overview

Peste des petits ruminants (PPR) is a devastating viral disease caused by the peste des petits ruminants virus (PPRV), a morbillivirus closely related to the rinderpest virus (RPV) and human measles virus (Njeumi *et al.*, 2020). Sheep and goats are the definitive hosts, with goats being more susceptible than sheep to PPRV (Diop *et al.*, 2005). Cattle, camelids and pigs are considered to be dead-end hosts (Schulz *et al.*, 2019). Morbidity rates in small ruminants can reach 100% (Parida *et al.*, 2015), with potential fatality rates of up to 90% (Singh *et al.*, 2009).

Peste des petits ruminants virus was first reported in Côte d'Ivoire but has now spread as far afield as Central Asia, China, South Asia and the European part of Turkey (Baron *et al.*, 2016). An estimated 60–80% of the world's farmed ruminant populations are threatened by PPRV (Baron *et al.*, 2016; Njeumi *et al.*, 2020). The recent outbreaks in Thrace and North Africa particularly threaten further spread within Europe (Parida *et al.*, 2015). However, PPRV has been targeted for eradication by 2030 by the World Organisation for Animal Health (WOAH) as many factors favouring the eradication of RPV are present in PPRV: reliable, safe and effective vaccine; simple and effective diagnostics; short infectious period with no carriers; no evidence of a wildlife reservoir; transmission by close contact only; short survival of PPRV in the environment; a single serotype of the virus (Njeumi *et al.*, 2020).

Prevalence

Peste des petits ruminants virus was first reported in Turkey in 1999, spreading to the regions bordering Greece in 2004 and Bulgaria in 2006. Outbreaks of PPRV in Turkish Thrace represent a serious threat to further spread into Europe (Altan *et al.*, 2019). The suspected spread in 2018 to Bulgarian villages bordering Turkey in 2018 presents a potential warning, although of the 7203 samples collected, only six were confirmed as seropositive with no evidence of virus circulation (SCoPAFF, 2018). A study in the Marmara region of Turkey (which borders Greece and Bulgaria) found PPRV antibodies in the sera of 11% of the sheep tested, in addition to observing PPR-like clinical signs in 59% of those samples (n=111). However, none of these were from border regions adjacent to Greece and Bulgaria (Altan *et al.*, 2019). The study also demonstrated that seroprevalence of PPRV antibodies in the Marmara region is lower than that reported elsewhere in Turkey, ranging from 14.9% to 87.9% (Tatar and Alkan, 1999; Ozkul *et al.*, 2002; Tatar *et al.*, 2002; Albayrak and Alkan, 2009). As of August 2024, PPR has been detected in both Greece and Romania, which were previously disease free (WOAH, 2024).

Risk factors

Risk factors for the introduction of PPRV to Europe include illegal movement of live animals from endemic countries; the movement of contaminated animal products; contamination of passenger clothing returning from endemic regions (Parida *et al.*, 2015); movement of incubating or infected wild cervids from endemic regions (Hoffmann *et al.*, 2012; Parida *et al.*, 2015); contaminated vehicles returning from endemic areas and/or improper cleaning of returning vehicles (Parida *et al.*, 2015).

Transmission

Transmission occurs by close contact with infected animals via exhaled aerosols, particularly due to dry coughing or through excretion of bodily fluids and waste such as faeces and urine. Spread may occur over large distances due to the movement of infected animals as part of normal trade or migratory patterns. There is a particular risk of introduction when animals are in the incubation phase without clinical signs (Parida *et al.*, 2015). PPRV can also spread to animals in a mild virulent form, which can then lead to severe disease at a later date (Couacy-Hymann *et al.*, 2007a). Transmission via fomites has also been posited (Parida *et al.*, 2015). Reintroduction or recurrence of PPR in areas where it has been thought to be eliminated have been attributed to the introduction of new live small ruminants incubating the disease, and illegal movements of animals and animal products. The recent reintroduction of PPR to Morocco in 2008 has been attributed to an intense trade in sheep and goats prior to Eid al-Adha (Parida *et al.*, 2015). Furthermore, there is growing concern that wild goats in Turkish Thrace may transmit PPRV to Greece and/or Bulgaria as a bridging species, as all three countries host interconnected communities of wild goats (Parida *et al.*, 2015), as in Kurdistan wild goats have been shown to exhibit clinical infection with PPRV (Hoffmann *et al.*, 2012).

Although pigs and cattle are dead-end hosts, there is some experimental evidence of pig–pig and pig–goat transmission (Schulz *et al.*, 2019). There is significant movement of live animals to endemic countries such as Morocco from Spain, France and Italy, and there is concern that transmission could occur via fomites in returning vehicles. PPRV can also survive for 2–3 days in meat, so there is a risk of transmission through this route (Miller, 2009).

Control measures and biosecurity

One of the most effective control measures for PPRV is vaccination, with commercially available live-attenuated vaccines readily available which generate long-lasting immunity of ~3 years (Sen *et al.*, 2010). These vaccines also offer cross-protection against the different genetic lineages of PPRV. With live-attenuated vaccines in rare cases reversion to a virulent form can occur, potentially leading to an outbreak, but there is significant ongoing research to develop additional vaccines using alternative technologies

(Rojas *et al.*, 2021a). However, no PPR vaccines are licensed within the EU (EFSA, 2021).

Regarding biosecurity, at the external level importation of animals from endemic countries into the EU is forbidden; however, returning trucks involved in the export of live animals to endemic countries must be properly cleaned prior to re-entry to the EU (Parida *et al.*, 2015). Additional effective measures to contain a potential outbreak include the creation of protection (3 km radius) and surveillance (10 km radius) zones, establishment of a monitoring period of 21 days and 33 days in establishments where first detected (EFSA, 2021). Containment of the outbreak in Bulgaria in 2018 effectively demonstrated these measures, The country further implemented an expanded surveillance zone of 20 km around the site of outbreak and undertook the culling of all susceptible animals within affected villages, sampling of all villages within the protection and surveillance zones, clinical examinations of all holdings with small ruminants in the border region with Turkey, multiple disinfections within affected villages and a ban on animal movements within and from the respective zones (SCoPAFF, 2018).

Availability of modelling data

A stochastic SEIR model of within-herd transmission exists wherein the basic reproduction number (R_0) and transmission rate (β) are provided (EFSA, 2021). Further estimations of R_0 are available for African contexts (Apolloni *et al.*, 2016; Hammami, 2016). Transmission rates and mortality (Zahur *et al.*, 2009; Kivaria *et al.*, 2013; Fournié *et al.*, 2018) and latent infection periods (Couacy-Hymann *et al.*, 2007b; Liu *et al.*, 2013; Parida *et al.*, 2019; Halecker *et al.*, 2020) are estimated based on experimental challenge publications. Further studies have examined the transboundary spread of PPRV in western China using a prediction model (Gao *et al.*, 2021), a dynamic transmission and elimination model in Ethiopia (Fournié *et al.*, 2018) and global ecological niche modelling of PPRV with an ensemble modelling algorithm (Assefa *et al.*, 2021). PPRV transmission scaling has been examined in the context of northern Tanzania (Herzog *et al.*, 2020).

Peste des petits ruminants virus can survive at 37°C for ~2 hours and is completely inactivated within 1 hour at 60°C, but can survive for long periods when frozen in tissue. It remains stable at pH 5.8–10 but is susceptible to most commonly used disinfectants including alcohol, ether, phenol and sodium hydroxide (Kumar *et al.*, 2014).

Foot and Mouth Disease

Overview

Foot and mouth disease (FMD) is a highly contagious disease of both wild and domestic even-toed ungulates caused by the foot and mouth disease virus (FMDV) (Brown *et al.*, 2022). FMD is characterized by fever, blisters and sores on the lips, tongue, mouth, teats and between hooves (WOAH). Although the mortality rate is typically ~5% (Longjam *et al.*, 2011), animals that do survive are often weakened and debilitated. FMD can also lead to severe myocardial lesions and death in young animals (Longjam *et al.*, 2011). It is endemic to parts of Africa, Asia and South America, where outbreaks are estimated to cause €6.7–19.61 billion in damages annually (Knight-Jones and Rushton, 2013). Sporadic outbreaks continue to occur in FMD-free countries, most famously the 2001 outbreak in the UK, which led to the destruction of 6.5 million animals and total losses of €9.32 billion (Brown *et al.*, 2022). Losses from FMD outbreaks typically extend far beyond direct losses from reduced production and changes in herd structure, with many indirect losses caused by restricted access to markets (Knight-Jones and Rushton, 2013), and can even include losses due to restrictions on tourism, as was the case in the UK outbreak of 2001 (Brown *et al.*, 2022).

Prevalence

At the time of writing, FMD has been eradicated from continental Europe, North America, Oceania (except Papua New Guinea), the majority of South America, Madagascar and parts of Botswana and Namibia. However, it persists in the majority of Asia and Africa, with countries neighbouring Europe such as Turkey and Morocco of particular concern (EFSA, 2021). Europe has experienced sporadic outbreaks, the most recent of which occurred in Bulgaria in 2011 (EFSA AHAW, 2012a). There are seven distinct serotypes of FMDV: O, A, C, SAT 1–3 and Asian-1, and with the exception of Asian-1, each serotype is endemic to its respective region, although serotypes have the capacity to spread and become established in new regions (Valarcher *et al.*, 2008; Lu *et al.*, 2022).

Movement of animals is considered to be the principal risk factor for the introduction of FMD into FMD-free territories (Martínez-López *et al.*, 2008). Delays in disease reporting constitute a significant risk factor for the introduction of FMD into a neighbouring country (Motta *et al.*, 2019), as was the case in 2001 with a lapse of a few days allowing FMD to spread from the UK to The Netherlands, Ireland and France (Chmitelin and Moutou, 2002), and outbreaks in Japan and South Korea in 2010 (Motta *et al.*, 2019). There is a risk of air-borne transmission, which can transport the virus short distances (2 km) to neighbouring farms or further afield, potentially to neighbouring countries up to 50 km over land or 200 km over water (Brown *et al.*, 2022). However, this risk will vary greatly depending on local geographic conditions, such as the species present, climate, density of animals and the use of different management systems (Miguel *et al.*, 2013; Brown *et al.*, 2022). The ideal climatic conditions, particularly moderate temperatures and higher humidity, are generally present in temperate countries in Europe (Miguel *et al.*, 2013).

Finally, pigs shed significantly more virus than ruminants and a high density of pigs in a given area has been identified as a risk factor for air-borne transmission of FMDV (Dion *et al.*, 2011; Klausner *et al.*, 2015; Brown *et al.*, 2022). In Europe, The Netherlands and Denmark are particularly high-risk regions due to their large pig-farming industries (Brown *et al.*, 2022).

Transmission

The principal transmission route for FMDV is by inhalation of virus through direct contact with the breath of infected animals (Paton *et al.*, 2018; EFSA, 2021). Air-borne transmission has been implicated, within both short (2 km) and long distances (up to 50 km over land and 200 km over water) (Brown *et al.*, 2022; Gloster *et al.*, 2003). Air-borne transmission of FMDV is generally posited when no other transmission route is thought possible in the circumstances, and has been implicated in a number of outbreaks (Brown *et al.*, 2022). Contact with fluids such as faeces, urine, saliva, milk and semen from infected animals can also transmit the virus (EFSA, 2021). FMDV is preserved by freezing and refrigeration, and survives in tissues that remain at neutral pH after death, including lymph nodes and bone marrow (WOAH Technical Disease Card), and can survive in moist (ideally humidity >55%) and cool temperatures (Brown *et al.*, 2022), including in contaminated feed and the environment for weeks to a month (WOAH Technical Disease Card).

Different species have varying susceptibility to FMDV, with ruminants requiring as little as 10 $TCID_{50}$ to initiate an infection (Kitching *et al.*, 2005). Cattle are thought to be most susceptible to air-borne infection due to their greater lung capacity (Alexandersen *et al.*, 2003a), whereas swine are estimated to require a much higher dose of 6000 $TCID_{50}$ (Alexandersen *et al.*, 2002). Conversely, despite swine being less susceptible to FMDV than cattle, they are an important source of aerosolized FMDV during outbreaks, and are capable of shedding 100–1000 times more virus than ruminants (Sellers and Parker, 1969; Donaldson *et al.*, 1982; Alexandersen *et al.*, 2002).

Control measures and biosecurity

One of the most important control measures preventing the introduction of FMD to an FMD-free territory is timely and accurate disease reporting, which requires that disease surveillance be undertaken, particularly in high-risk regions bordering endemic countries, as is the case in Thrace (Motta *et al.*, 2019). Active surveillance consists of frequent and intensive assessment of the disease presence in a given region/population (Motta *et al.*, 2019). Passive reporting is defined as observer-initiated surveillance, such as voluntary notification based on clinical signs by animal health professionals, farmers and associated farm workers/advisors/processors, etc. (Hoinville *et al.*, 2013; Motta *et al.*, 2019). In line with this, the transboundary high-risk area epidemio-(THRACE) surveillance programme conducts targeted active surveillance assessing clinical signs, serosurveillance and abattoir surveillance in the Thrace region (EFSA AHAW, 2012b; Motta *et al.*, 2019). In addition to active and passive surveillance, establishment of protection zones with a minimum radius of 3km, and surveillance zones with a minimum radius of 10km are considered to be effective (EFSA, 2021).

Finally, a minimum 21-day waiting period following final cleansing and disinfection is mandated by the EU directive prior to restocking (EFSA, 2021), due to the potential persistence of FMDV in the environment, potentially up to 14 days in the correct environmental conditions (Colenutt *et al.*, 2020).

Although a commercial vaccine for FMD is available, with inactivated virus (Lu *et al.*, 2022), this is no longer employed prophylactically in the EU, with only emergency vaccination in the case of an outbreak being used at present (Lu *et al.*, 2022). There are some drawbacks to the current inactivated FMDV vaccines: certain serotypes are more unstable than others; whole virus particles can quickly degrade into less immunogenic subunit particles; and there is a lack of cross-protection between certain serotypes necessitating the use of pandemic strains (Lu *et al.*, 2022). Thus, effective biosecurity and surveillance are essential in preventing the reintroduction of FMD to free regions.

Availability of modelling data

A simple generic model of FMD has been constructed where the between-farm basic reproductive rate is calculated (Thornley and France, 2009), and there has been a comparison of models to capture the potential for disease transmission in sheep flocks (Schley *et al.*, 2012). EuFMDiS is a continental-scale transboundary model of livestock disease to support emergency animal disease planning in Europe, with a focus on FMD (Bradhurst *et al.*, 2022). Howey *et al.* (2012) modelled the within-host dynamics of FMDV in cattle in experimental infection.

Transmission via a contaminated environment has been examined experimentally in cattle, with R_0 estimated at 1.65 (Colenutt *et al.*, 2020), and in calves, with R_0 estimated at 1.9 (Bravo de Rueda *et al.*, 2015), suggesting that an outbreak may be sustained by this transmission route alone. Experimental infection studies examining aerosol transmission have been conducted for cattle (Pacheco *et al.*, 2010) and sheep (Stenfeldt *et al.*, 2015, 2019). A review of intranasal infection of cattle, sheep and pigs covers the infectious doses for each species (Sellers and Gloster, 2008). Further studies have examined between-pen transmission from pigs to cattle (Colenutt *et al.*, 2016) and pigs to pigs (Eblé *et al.*, 2006; Pacheco *et al.*, 2010). The incubation period for FMDV has been studied in piglets, lambs, calves, and dairy cows (Orsel *et al.*, 2009), sheep and goats (Kitching and Hughes, 2002). Carpenter *et al.* (2004) constructed a simulation model of intra-herd FMD transmission. Farm-to-farm transmission and the role of surveillance have been studied and a between-farm reproductive rate calculated (Nelson *et al.*, 2017). Within-farm transmission has been examined in the context of the 2001 UK outbreak (Chis Ster *et al.*, 2012).

A number of studies have examined experimental infection using indirect transmission experimental design in swine (Alexandersen and Donaldson, 2002; Alexandersen *et al.*, 2002), sheep (Gibson and Donaldson, 1986), and swine and cattle (Alexandersen *et al.*, 2003b). Ward *et al.* (2015) modelled FMD transmission in a wild pig and domestic cattle ecosystem. The EFSA has modelled transmission in the context of the design of control measures for containing outbreaks (EFSA, 2021), with an earlier EFSA report also focusing specifically on Thrace (AHAW, 2012b). Epidemic simulations of FMD outbreak in Minnesota, USA, have been modelled (Gale *et al.*, 2015). The re-aerosolization of FMDV from sources such as fomites or skin through daily husbandry practices has been examined (Dillon, 2011), while earlier studies developed epidemic simulation models for FMD control strategies (Bates *et al.*, 2003).

Several studies have examined air-borne and wind-borne transmission, Klausner *et al.* (2015) used a Lagrangian stochastic model to predict the required density (1000/km^2) of pigs for successful air-borne transmission. Björnham *et al.* (2020) constructed a multi-level model for air-borne transmission of FMD in Swedish livestock. Schley *et al.* (2009) modelled air-borne transmission of FMD from a known infected premises. Gloster *et al.* (2010) carried out an air-borne transmission model intercomparison, while Seo *et al.* (2015) used computational fluid dynamics to model air-borne spread. Further studies have modelled the introduction of FMD into Australia (Garner *et al.*, 2006; Fountain *et al.*, 2018; Hernández-Jover *et al.*, 2016), New Zealand (Heuer *et al.*, 2007) and Alberta, Canada (Sanson *et al.*, 2014), the spatial dynamics of spread in Australia in feral pigs and livestock using a susceptible–infected–recovered automata model (Doran and Laffan, 2005), a transmission network model examining Japan and Australian contexts (Firestone *et al.*, 2020) and spread across the USA (Hagerman *et al.*, 2018; Coffman *et al.*, 2021).

Sellman *et al.* (2020) examined realistic assumptions of spatial locations and clustering of premises for FMD disease spread models in the USA. Souley Kouato *et al.* (2018) reviewed epidemiological risk models for FMD with a focus on Africa. Tsao *et al.* (2020) considered the differences in regions and demography on modelling FMD in cattle at the national scale. A further study examined the role of farm contiguity and its implications for FMD in Scotland (Flood *et al.*, 2013). Donaldson *et al.* (1982) used prediction models to forecast and analyse air-borne spread during outbreaks in the Channel Islands and Brittany, France. Keeling (2005) summarized and compared the three models used in the 2001 UK FMD outbreak. Tildesley and Keeling (2008) compared modelling FMD in the UK and Denmark. Temporal-spatial analysis of age-space structured FMD has been modelled with a Dirichlet boundary condition (Wang *et al.*, 2021), and temporal investigations of reported movements and demographics of cattle and pigs within Sweden (Nöremark *et al.*, 2009).

Studies have also examined the initial spread of FMD through animal movements (Green *et al.*, 2006), the role of cattle transport networks in endemic and epidemic FMD risk in Turkey (Herrera-Diestra *et al.*, 2022), the role and demographic structure of livestock movements in the UK (Kao *et al.*, 2006), the role of wild ungulates in the potential spread of FMD in Thrace (Dhollander *et al.*, 2016), and the role of ship and aircraft waste in the potential introduction of FMD into Scotland (Adkin *et al.*, 2008). Mass vaccination and population-level control and protection in Turkish cattle (Knight-Jones *et al.*, 2016), emergency vaccination during an FMD outbreak in Minnesota, USA (Miller *et al.*, 2015), and epidemiological and economic considerations for vaccination strategies in FMD-free countries (Yadav *et al.*, 2022) have also been modelled.

In terms of pathogen survival, FMDV can survive in temperatures up to 50°C, but is progressively inactivated in temperatures of >50°C–<70°C, and is fully inactivated by heating to >70°C for >30 min, pH <6.0 or >9.0, and by the following disinfectants: 2% sodium hydroxide; 4% sodium carbonate; 0.2% citric acid; 2% acetic acid; 3% sodium hypochlorite; 1% potassium peroxymonosulphate/sodium chloride; and chlorine dioxide. FMDV is resistant to iodophores, quaternary ammonium compounds and phenol particularly in the presence of organic matter (WOAH Technical Disease Card). Furthermore, experimental studies have examined the stability of different FMDV serotypes in aerosols (Barlow, 1972; Brown *et al.*, 2021) and cell culture media vs bovine saliva (Barlow and Donaldson, 1973).

Avian Influenza

Overview

Avian influenza viruses (AIVs) belong to the *Influenza A* genus, with their ancestral origin and primordial reservoir in aquatic birds (Lee *et al.*, 2021). AIVs can be divided into low-pathogenicity (LPAI) and high-pathogenicity (HPAI) avian influenza, and further categorized by the 16 haemagglutinin (H) and nine neuraminidase (N) subtypes (Swayne *et al.*, 2020), with H5 and H7 being responsible for the most deadly and systemic disease (Lee *et al.*, 2021). Since the late 1950s, the majority of HPAI outbreaks have been attributed to the virulence shifts in LPAIs, which have had limited spread and been largely eradicated (Lee *et al.*, 2021). The current ongoing outbreak across Asia, Africa, Europe and North America involves the A/goose/Guangdong/1/1996 lineage of H5 (Swayne *et al.*, 2020). This outbreak has led to the culling of 77 million birds to curb the spread of the virus globally (Miller, 2022), causing enormous economic impact (Hunter, 2022). In Europe, there have been thousands of recorded outbreaks to date, with upwards of 46 million birds culled (Hunter, 2022; EFSA, 2023).

Prevalence

From December 2022 to March 2023, there were a total of 6615 HPAI detections in Europe across 37 countries, with the virus spreading to seabirds in addition to captive birds and poultry. HPAI outbreaks in poultry were recorded in France (191), Poland (88), Hungary (31), Czechia (26), Germany (20), UK (19), Belgium (6), Austria, Denmark, The Netherlands, Romania, Slovakia (3 each), Italy, Spain (2 each), Bulgaria, Moldova and Slovenia (1 each). A significant portion of the outbreaks recorded in France, Poland, Hungary, Romania, Belgium and, to a lesser extent Czechia were secondary outbreaks (EFSA, 2023), having spread from other poultry farms.

Risk factors

Given the role that wild waterfowl play in disseminating avian influenza globally, one of the principal risk factors for transmission of the virus into the commercial poultry industry is the proximity of a water source, as this may lead to close interaction between migratory waterfowl and domesticated birds such as ducks or chickens (Chatziprodromidou *et al.*, 2018). Viral shedding by infected birds can also contaminate poultry worker clothing and farm equipment, so a lack of proper disinfection and biosecurity measures to decontaminate facilities and employees is also a significant risk factor, and the primary cause of dissemination within commercial poultry flocks (Simancas-Racines *et al.*, 2023). Finally, the presence of pigs in close proximity to wild and domesticated birds constitutes a significant risk factor for the reassortment and emergence of novel avian influenza strains, as pigs are considered a likely mixing vessel for certain strains (Su *et al.*, 2015).

Transmission

Bird–bird transmission can occur through a number of pathways, depending on whether the infection is spreading between wild birds, from wild to domesticated, or between domesticated birds (Simancas-Racines *et al.*, 2023). Waterfowl constitute the natural reservoir of avian influenza and play a role in spreading the virus globally through long-distance migration (Su *et al.*, 2015). Wild waterfowl can transmit the infection to wild land birds and domesticated birds, via contaminated water sources and/or food sources (Yamaji *et al.*, 2020; Guinat *et al.*, 2022). This is of particular importance where backyard farming systems exist (Su *et al.*, 2015).

The faecal–oral route remains the most significant transmission route, as infected birds shed significant quantities of virus in faeces, and avian influenza can be transmissible for up to 21 days (Wahlgren, 2011; Bui *et al.*, 2016). Transmission can also occur via bodily secretions such as mucus, saliva and urine, which contaminate clothing and footwear of poultry workers, cages and equipment (including for egg collection) (Simancas-Racines *et al.*, 2023). This route is considered the primary transmission route within commercial flocks (Simancas-Racines *et al.*, 2023).

Finally, avian influenza virus is resistant to temperatures below 28°C and can survive for up to 200 days in bodily fluids of infected birds, 4 days in faeces at physiological temperatures, 35 days at 4°C, in carcasses, meat and eggs, and up to 5 weeks in the poultry house environment (Blagodatski *et al.*, 2021).

Control measures and biosecurity

Aside from the standard recommended biosecurity protocols commonly applied for the poultry industry, including proper disinfection of equipment and clothing, not bringing personal items into hatchery rooms, limiting the number of farms visited by veterinarians and other visitors per day, employing an all-in/all-out system, and preventing contacts between farms (Ssematimba *et al.*, 2013), there are a number of measures specifically recommended for a suspected avian influenza case to prevent further spread leading to a large-scale outbreak.

First, a containment protocol should be followed and a clinical diagnosis carried out by trained personnel and if possible, sample collection for dispatch to specialist laboratories. Then, creation of a register of farms, poultry and domestic fowl within the suspected outbreak area, opening of communication channels between stakeholders and relevant animal health authorities, and reports sent to international institutions. If the case is confirmed, poultry records should be generated for the area, including backyard farms, and an epidemiological investigation opened to determine the transmission source and prevent new contact cases if possible. This involves the creation of an infection zone within 1 km of case zero where drastic measures such as depopulation and sanitary culling, the elimination of carcasses, products and by-products, and strict cleaning and disinfection are undertaken. An observation zone within 3 km of case zero should be observed, where transport of poultry and poultry products is forbidden. Finally, a surveillance zone should be instituted within 7–10 km of case zero where markets selling poultry and poultry products should be closed to prevent further spread (Simancas-Racines *et al.*, 2023).

Availability of modelling data

Kirkeby and Ward (2022) have compiled a review of transmission parameters for the spread of avian influenza, covering the poultry species, study unit of transmission (within/between flock), estimation period, transmission rate (b), R_0, latency in days and period of infectiousness (a). Furthermore, these authors estimate R_0 in studies where a and b are available. This generated a total of 100 R_0 estimates, 40 for H5N1, 16 for H5N2, 15 for H7N1, 11 for H7N7, 7 for H7N3, and one for H5N3, with a further six not specifying the serotype. The bulk of estimates are based on chickens (56), with poultry (28), turkeys (13) and ducks (3) making up the remainder. In addition, 28 estimates of latency period and 45 estimates of period of infectiousness are reported.

Beyond this extensive review, a number of studies have modelled various other aspects of transmission including spatial spread through farms in Italy (Dorigatti *et al.*, 2010), wind-borne spread during the 2003 outbreak in The Netherlands (Ssematimba *et al.*, 2013), within-flock transmission in France (Vergne *et al.*, 2021), a review of the risk factor modelling for spatio-temporal patterns of spread (Gilbert and Pfeiffer, 2012), a review of the role of animal and human movements in avian influenza spread (Hautefeuille *et al.*, 2020), and mathematical modelling of treatment and vaccination for poultry farms (Malek and Hoque, 2022). Several studies have considered the survival of avian influenza virus on human skin and in the presence of ethanol (Bandou *et al.*, 2022), the inactivation of the virus by common disinfectants (Lombardi *et al.*, 2008), and the inactivation of the virus by chemical agents and under various physical conditions (De Benedictis *et al.*, 2007).

Finally, economic analysis has examined HPAI control in The Netherlands (Longworth *et al.*, 2014), and a global systematic review of avian influenza outbreaks between 2010 and 2016 by distribution, avian species and virus strain is available (Chatziprodromidou *et al.*, 2018).

Sheep and Goat Pox

Overview

Sheep pox (SPP) and goat pox (GTP) are viral diseases caused by capripoxviruses infecting sheep and goats, and are closely related to the lumpy skin disease (LSD) virus affecting cattle (Tuppurainen *et al.*, 2017). SPP and GTP are characterized by fever, generalized papules and in some cases vesicles, and the potential for internal lesions and death. They are economically important diseases of sheep and goats in endemic regions, particularly in western Asia and Africa where they cause high morbidity and mortality (Zewdie *et al.*, 2021). Animal losses and damage to productivity can take years to recover from and lead to annual losses of 30–40% (Garner and Lack, 1995). The recent conflict in Syria and the displacement of millions of refugees to surrounding countries has led to a concurrent increase in the movement of livestock. Many refugees have travelled to border regions with potentially unvaccinated small ruminants, which has been associated with the current spread of SPP and GTP (Tuppurainen *et al.*, 2017). Although not widespread within Europe, there have been increasing reports of outbreaks in countries neighbouring Europe, including in the far east of the Russian Federation (Maksyutov *et al.*, 2015), and in countries in North Africa (Kardjadj, 2017; Ben Chehida *et al.*, 2018).

Prevalence

Although prevalent in many regions of the world, SPP and GTP are largely absent from Europe and North America (Tuppurainen *et al.*, 2017; Zewdie *et al.*, 2021). Cases regularly occur across much of Africa and western Asia, particularly North African countries neighbouring Europe, including Tunisia and Algeria (Kardjadj, 2017; Ben Chehida *et al.*, 2018). SPP and GTP are endemic in Turkey and there have been outbreaks in Bulgaria and Greece (Mangana-Vougiouka *et al.*, 2008; Malesios *et al.*, 2014; Tuppurainen *et al.*, 2017). SPP and GTP outbreaks have occurred recently in Cyprus and Spain (DEFRA, 2023).

Risk factors

The principal risk factor for the introduction of SPP or GTP into Europe is the unregulated movement of infected animals from endemic areas (Mangana-Vougiouka *et al.*, 2008; Tuppurainen *et al.*, 2017). Lack of compulsory and/or consistent vaccination is also a cause for concern in endemic regions neighbouring Europe, particularly as nomadic and transhumance-based practices can transport animals over long distances (Tuppurainen *et al.*, 2017). There is also a risk in some endemic areas of the use of expired or otherwise unauthorized vaccines with unknown viral titres that may not offer suitable levels of protection (Abutarbush, 2014).

Transmission

Experimental studies have shown that sheep and goats can be infected by SPPV and GTPV respectively via intradermal, intranasal and intraoral routes. Viral shedding can occur via oral, nasal and ocular secretion, with transmission occurring via aerosols and/or direct contact with an infected animal (Bowden *et al.*, 2008). SPPV and GTPV are very stable in the environment and can persist for long periods of time, which can lead to the infection of naïve animals (Tuppurainen *et al.*, 2017). Infected animals can also exhibit high viral loads on the skin, and experimental studies have shown that certain insect vectors can potentially transmit the virus mechanically (Kitching and Mellor, 1986). However, their role in the field is not well understood. The lack of seasonal transmission patterns mirroring the population dynamics of the vector further suggests that it is unlikely that this represents a major transmission pathway (Tuppurainen *et al.*, 2017).

Control measures and biosecurity

Effective and cheap (€1.50–2.00) live attenuated vaccines are available for SPPV and GTPV, and can provide good protection if a herd immunity of ~80% is maintained by annual vaccination (Tuppurainen *et al.*, 2017). Outbreaks can be successfully controlled with timely vaccination (Ben-Gera *et al.*, 2015). Successful control and eradication are heavily dependent on the early detection of initial cases and the quick implementation of quarantine, disinfection, the removal of all infected and in-contact animals and the strict imposition of movement controls in the affected area. These measures require good levels of awareness amongst veterinarians and farmers, in addition to diagnostic capacity (Tuppurainen *et al.*, 2017).

Availability of modelling data

Several studies have examined the epidemiology of SPP and GTP in Greece, analysing historical data from outbreaks in the 1980s and 1990s to the late 2000s (Mangana-Vougiouka *et al.*, 2008; Malesios *et al.*, 2014). Experimental infections have also been carried out to determine transmission routes (Kitching and Taylor, 1985; Kitching and Mellor, 1986; Bowden *et al.*, 2008; Hamdi *et al.*, 2021).

Classical Swine Fever

Overview

Classical swine fever (CSF) is a viral disease of pigs and boars caused by the classical swine fever virus (CSFV) (Blome *et al.*, 2017), a flavivirus closely related to bovine viral diarrhoea virus and several other viruses of veterinary importance (Moennig, 2015; Fan *et al.*, 2021). CSF is highly contagious and a cause of significant morbidity and mortality, and economic impact to the pig industry globally (Fan *et al.*, 2021). Although largely eradicated from the European Union at the time of writing, previous outbreaks in EU countries have caused enormous losses. The 1997 outbreak in The Netherlands alone is estimated to have cost upwards of €2.1 billion, mostly due to destruction of swine as part of preventtive measures to contain the spread (Moennig, 2015).

Classical swine fever virus causes three distinct forms of disease: acute, which can be transient or lethal; chronic; and persistent which typically occurs when the animal is infected prenatally (Moennig *et al.*, 2003). Incubation is typically 4–7 (rarely 10) days post infection, and progression is dependent on the strain virulence, host response and secondary infections (Blome *et al.*, 2017). The acute form of CSF disease typically leads to unspecific clinical signs such as fever, anorexia, general weakness, conjunctivitis and gastrointestinal distress (Petrov *et al.*, 2014). This can progress to neurological symptoms 2–4 weeks after infection, often with skin haemorrhages or cyanosis, which are characterized as CSF-specific symptoms. Mortality rates can reach 100% depending on the strain. CSF also causes suppression of the host's immune system, which can lead to devastating secondary infections, further complicating the disease course. Chronically infected animals, however, only display non-specific symptoms but shed high amounts of the virus prior to their eventual death within a month of infection. Persistent infection typically occurs when the sow is infected between days 50 and 70 of pregnancy, with piglets born apparently healthy but succumbing several months later to late-onset CSF. During the intervening period they shed significant amounts of virus (Blome *et al.*, 2017). Together with persistent infection, this can have a significant effect on transmission within wild boar populations (Kaden *et al.*, 2005).

Prevalence

Classical swine fever has been eradicated from the EU, North America, Argentina, Australia and New Zealand (ESFA AHAW, 2021). However, it is endemic in many parts of the world such as China, Russia, South-East Asia and much of South America and the Caribbean, whereas much of Africa has not certified status regarding CSF (Postel *et al.*, 2017). There have been sporadic outbreaks in EU territories within the last decade, Lithuania in 2009 and 2011 and Latvia in 2012–2015 for example (Postel *et al.*, 2013). Within the EU, much of the wild boar population is also thought to be free of CSF (Postel *et al.*, 2017).

Risk factors

One of the principal risk factors for the introduction of CSF is the transport and trade of pigs and pig products, in addition to the overall density of pigs in an area (Postel *et al.*, 2017). This has been extensively explored in a number of studies, often focusing on the outbreaks of CSF in The Netherlands in the mid to late 1990s (Stegeman *et al.*, 1999; Boender *et al.*, 2014). Containment and control of an outbreak of CSF are heavily dependent on early detection, which can be further complicated by the emergence of lower virulence strains in recent years which can mask the clinical signs of CSF, delaying detection (Postel *et al.*, 2017). Finally, endemic CSF in wild boar is also a significant risk factor, being further compounded by the potential contact between boar in CSF-free countries and endemic countries. This is of particular concern where there is the combination of dense boar populations, dense domestic pig populations and close contact between wild boar and domestic pigs (Postel *et al.*, 2017).

Transmission

Hosts for CSFV are members of the Suidae family, principally domestic pigs and wild boar. CSF can be transmitted both directly and indirectly, and vertically and horizontally (Blome *et al.*, 2017). Indirect transmission typically takes place via feeding of virus-contaminated swill and mechanical transmission through contact with humans or veterinary/farm equipment (Van Oirschot, 1999; Ribbens *et al.*, 2004), while some experimental studies have suggested a potential role for air-borne transmission (Laevens *et al.*, 1999; Weesendorp *et al.*, 2008, 2009). Direct transmission often occurs via the oronasal route and less often via conjunctiva, mucous membranes, skin abrasions, sexual or artificial insemination and contaminated instruments (Blome *et al.*, 2017). Virus is typically shed in saliva, mucus, lacrima, urine, faeces and semen (Van Oirschot, 1999; Ribbens *et al.*, 2004; Blome *et al.*, 2017). Vertical transmission can occur at any stage during gestation, but the production of persistently infected piglets, rather than stillbirths, requires infection between days 50 and 70 of pregnancy (Blome *et al.*, 2017). Transmission from wild boar to domestic pigs is also a significant route, with up to 60% of all primary CSF outbreaks in Germany in the period 1993–1998 traceable back to the wild boar reservoir (Fritzemeier *et al.*, 2000).

Control measures and biosecurity

Classical swine fever vaccines are typically derived from live-attenuated vaccine strains (Fan *et al.*, 2021), and are a mainstay of mitigation strategies in a number of countries, including the Russian Federation, China and many South-East Asian countries (Postel *et al.*, 2017). However, prophylactic vaccination is prohibited in many CSF-free countries, including the EU, mainly as these vaccines lack serological distinction between infected and vaccinated animals (Postel *et al.*, 2017; Fan *et al.*, 2021). Eradication programmes are costly and require significant investment in adjunct strategies in addition to vaccination (Postel *et al.*, 2017), including movement restrictions, as were employed in Canada, and the combination of vaccination with stamping-out policies which led to the destruction of large numbers of animals in the USA, which may no longer be practicable due to ethical concerns with this strategy (Coronado *et al.*, 2021).

Current guidelines following the detection of a case of CSF within the EU recommend the establishment of a restriction zone, where movement of animals in and out is prohibited, and a protection zone with a radius of up to 3 km, official veterinarian visits to all non-affected establishments within the protection zone at least once, and extensive sampling to rule out the further spread of CSF. A surveillance zone must also be established in which clinical examination and sampling are conducted in a sample of the establishments within the zone (EFSA, 2021).

Availability of modelling data

A recent EFSA AHAW report compiled a number of transmission parameters for CSF in Europe, with various values supplied for low, medium and high transmission rates, short and long infectious periods, in addition to the latent period and mean infectious period (EFSA, 2021). Further studies have examined both between-farm and within-farm transmission, through direct contacts such as animal movements and indirect contacts such as through vehicle and person contacts, based on real data from Segovia, Spain (Martínez-López *et al.*, 2012) and a Be-FAST model for between-farm spread (Ivorra *et al.*, 2014). Within- and between-pen transmission and the calculation of R_0 from transmission experiments have been carried out in several studies (Klinkenberg *et al.*, 2002; Durand *et al.*, 2009). Both β and R_0 have been estimated for wild pigs (Hone *et al.*, 1992). Air-borne CSFV transmission within and between pens has been examined, and β estimated (Weesendorp *et al.*, 2014). Values for R_0 and β have been reported during experimental infections based on different strains of CSFV including the Paderborn, Zoelen and Brescia strains (Weesendorp *et al.*, 2009). Studies have considered the role of between-farm distance and farm size (Boender *et al.*, 2014), and the quantification of between-herd transmission (Stegeman *et al.*, 1999) during the 1997–1998 outbreak in The Netherlands.

Swine Vesicular Disease

Overview

Swine vesicular disease (SVD) is a viral disease of domestic pigs caused by the swine vesicular disease virus (SVDV) of the family Picornaviridae. All isolates fall under one serotype. SVD does not show rapid spread, often being restricted to single pens, and has low mortality, morbidity and speed of spread. It is thus considered low priority due to the limited production losses, frequent subclinical nature of the infection and negligible mortality and capacity to spread rapidly (EFSA AHAW, 2012a).

Prevalence

Within Europe, the disease is largely restricted to Italy, with occasional outbreaks reported in Portugal (2003, 2007) (EFSA AHAW, 2012a).

Risk factors

Swine vesicular disease virus is resistant to normal disinfectants and persists in the environment (EFSA AHAW, 2012a).

Transmission

Direct contact between susceptible and infected animals or the contaminated environment is in most cases necessary for spread. SVD spread when introduced into a naïve farm is often limited, and infection is restricted usually to a few pens. Between-farm spread is almost entirely related to animal movements and/or contaminated transport vehicles and personnel moving between farms. Swill feeding is a minor transmission route, accounting for a maximum of 15% of cases (EFSA AHAW, 2012a).

Control measures and biosecurity

Good standard pig production biosecurity can easily prevent SVD infections occurring and limit spread if they do (EFSA AHAW, 2012a).

Availability of modelling data

EFSA AHAW (2012a) modelled within-farm and between-farm transmission for SVD, and the transmission parameters are detailed therein.

Bluetongue

Overview

Bluetongue is a vector-borne viral disease of domestic ruminants and wild ungulates, caused by the bluetongue virus (BTV) (Courtejoie *et al.*, 2018). BTV has 27 known serotypes (Zientara *et al.*, 2014), with disease outcome varying based on the serotype and the species of animal involved (Mellor and Boorman, 1995). Although all ruminants are susceptible, the disease is of greatest concern in sheep (Rushton and Lyons, 2015). This can vary significantly from asymptomatic to fever, oedema, abortions and potentially death (Courtejoie *et al.*, 2018). It is estimated that BTV causes damage upwards of €2.73 billion, but the vast majority of this impact is due to restrictions to combat its spread in epidemic situations, which has had a far greater impact than productivity losses in endemic regions (Rushton and Lyons, 2015).

Prevalence

Of the 27 serotypes of BTV, only BTV-1, 2, 4, 8, 9, 16 and the small ruminant adapted strains 25, 27 and 30 are currently circulating in Europe. There have been periodic outbreaks of certain serotypes in different regions, such as the BTV-4 outbreak in the Balkans in 2014, which were BTV free for years, and resurgences, for example BTV-8 in France in 2015 (EFSA AHAW, 2017a). The capacity for an outbreak of BTV is, however, entirely dependent on the presence of the disease vector, mainly *Culicoides* spp. (Courtejoie *et al.*, 2018). As such, there are three distinct episystems for the Europe region: south-western, including the western Mediterranean, south-eastern encompassing the Balkans, and the northern and central European, characterized by primarily cattle-related *Culicoides* spp. and the least favourable climatic conditions (EFSA AHAW, 2017a). Prior to the summer of 2006, BTV had not been detected in northern Europe (>50 °N) when BTV-8 outbreaks were first reported in The Netherlands, Belgium, Germany, France and Luxembourg, with increasing spread in the following years (Wilson and Mellor, 2008, 2009).

Risk factors

The single most significant risk factor for BTV is the presence of the disease vector (EFSA AHAW, 2017a; Courtejoie *et al.*, 2018). In line with this, high temperature, particularly for the northern European episystem, is also a significant risk factor for the spread of the disease vector (Gubbins *et al.*, 2008), its survival and activity level (Courtejoie *et al.*, 2018), with a threshold temperature of 15°C (Boender *et al.*, 2014) below which an epidemic would fade out again (Gubbins *et al.*, 2008; Boender *et al.*, 2014; Courtejoie *et al.*, 2018). Furthermore, some strains are higher risk for increased pathogenicity than others, such as some BTV-2, BTV-4 and BTV-8 variants, and other strains have the potential to spread to other episystems (EFSA AHAW, 2017a). There is also a risk of horizontal transmission through contaminated needles (Rojas *et al.*, 2021b).

Transmission

Bluetongue virus transmission to ruminants is principally through the bite of *Culicoides* spp. (Baylis *et al.*, 2008). Other transmission routes are also possible, although rarer; for example, carnivores can in rare cases become infected after feeding on the meat of infected ruminants or wild ungulates (Jauniaux *et al.*, 2008). There is experimental evidence of horizontal transmission of BTV-1, 2 and 8 to naïve animals when housed together (van der Sluijs *et al.*, 2011, 2013; Rasmussen *et al.*, 2013), with the oral route being posited as likely due to shared water troughs and food (Rojas *et al.*, 2021b). Ingestion of contaminated placenta and colostrum has also been suggested as a potential horizontal transmission route in the field (Menzies *et al.*, 2008; Mayo *et al.*, 2010). However, this route of transmission is unlikely to play a significant role epidemiologically (Rojas *et al.*, 2021b).

Beyond vector transmission, vertical transmission is the most significant alternative route with the potential for venereal transmission leading to subsequent vertical transmission (De Clercq *et al.*, 2021). However, vertical transmission is heavily isolate dependent (Rojas *et al.*, 2021a).

Control measures and biosecurity

Control measures generally centre on the use of vaccination to eradicate the outbreak, and infection prevalence surveillance of ruminant populations using RT-PCR or c-ELISA (EFSA AHAW, 2017b)

Availability of modelling data

A systematic review was carried out for mathematical models of BTV prior to 2018, with R_0 estimates provided for several epidemics in Europe, such as The Netherlands outbreak of 2007, and a number of studies showing R_0 for within herds, between herds and an analysis of the impact of movement restrictions, vaccination and control measures (Courtejoie *et al.*, 2018). A network model has been constructed modelling bluetongue spread in France (Courtejoie *et al.*, 2019), and a cost analysis of BTV-8 surveillance was carried out (Hadorn *et al.*, 2009; Rüegg *et al.*, 2018). Risk models for BTV have been produced for England (Grace *et al.*, 2020) and outside Europe, including Kazakhstan (Abdrakhmanov *et al.*, 2021).

The dynamics of biting midges has been linked with climate data in a multiscale model focusing on the northern European episystem (Möhlmann *et al.*, 2021). Quantitative analysis of spatial and temporal evolution of BTV has been carried out on Lesvos, Greece (Malesios *et al.*, 2020) and in France (Charron *et al.*, 2013) and England with reference to both vector and animal movements (Turner *et al.*, 2012), and Sicily for vector movements (Purse *et al.*, 2004). Models have also analysed the risk of introduction to Australia (Eagles *et al.*, 2014). A thermodynamic and mechanistic model has been constructed for temperature and competence of the BTV vector (Gale, 2019).

Lumpy Skin Disease

Overview

Lumpy skin disease (LSD) has a serious impact on the beef industry worldwide, with the potential for up to 10% mortality (Whittle *et al.*, 2023). LSD typically presents with fever, nasal and ocular discharge and the characteristic nodular lesions on the skin, in addition to muscles and mucosal tissue (Möller *et al.*, 2019). In addition to high mortality, LSD leads to productivity losses for both meat and milk, fertility issues, damage to hides and secondary bacterial infections in some cases, with resulting trade restrictions and vaccination campaigns further compounding economic losses (Tuppurainen *et al.*, 2021; Whittle *et al.*, 2023). Lumpy skin disease virus (LSDV) is a member of the Poxviridae family, closely related to SPPV and GTPV. This close relationship led to the use of a combination vaccine containing a mix of live-attenuated Capripox vaccines in Eastern Europe, which caused a series of recombination events and the emergence of a virulent LSDV which spread throughout Asia (Whittle *et al.*, 2023).

Prevalence

Until recently, LSD was endemic to, and confined to, Africa, but in recent years it has spread to adjacent countries, initially in western Asia between 1989 and 2011 (Vandenbussche *et al.*, 2022; Whittle *et al.*, 2023), and to Europe itself in the Balkans (Agianniotaki *et al.*, 2017; Manić *et al.*, 2019). These outbreaks have, however, been contained with vaccination (Klement *et al.*, 2020; Whittle *et al.*, 2023), but there is concern for extension of the disease due to its rapid spread in adjacent regions and to Asia (Whittle *et al.*, 2023). There are ongoing outbreaks in Russia (Sprygin *et al.*, 2018) and Turkey (EFSA, 2020) which border Europe.

Risk factors

The ongoing outbreak in Turkey (EFSA, 2020) and the recent spread of LSD to the Balkans and across Asia (Whittle *et al.*, 2023) constitute the principal risk to further spread into Europe. The wide range of vector species capable of transmitting LSD, coupled with the capacity of the virus to infect wild bovids, in addition to domestic cattle, and the increased vector range are further compounding risk factors (Whittle *et al.*, 2023).

Transmission

The principal transmission route for LSD is via a variety of arthropod vectors, such as ticks, mosquitoes and biting flies (Whittle *et al.*, 2023). Some experimental evidence also suggests mechanical transmission by *Stomoxys calcitrans* and *Haematopota* spp. may be possible (EFSA, 2020). Climate change has the capacity to further spread LSD, with rising temperatures expanding the range of the disease vectors (Saegerman *et al.*, 2021). Although not initially considered an important route for transmission, recent evidence suggests, at least for the virulent recombinant Saratov/2017 strain, that infection via direct contact is possible (Aleksandr *et al.*, 2020), and possible transmission via fomites or overwintering in cattle has also been posited (Shumilova *et al.*, 2022). However, further research is needed to clarify this (Whittle *et al.*, 2023). Seminal transmission following artificial insemination has also been shown to be possible in experiments (Annandale *et al.*, 2014).

Control measures and biosecurity

The principal method of control for LSD is vaccination with live-attenuated vaccines, which has been effective in controlling the recent Balkan outbreak (Klement *et al.*, 2020). One important aspect of control is the disinsectization of animal transport vehicles to prevent further spread of LSDV (Saegerman *et al.*, 2021). LSDV is highly stable in the environment and can persist for long periods in necrotic lesions, desiccated crusts and dried hides (Namazi and Khodakaram Tafti, 2021). Turkey has been successfully reducing the incidence of LSD by adopting the following measures: mandatory mass vaccination free of charge; selective culling with 100% compensation; and quarantine measures (EFSA, 2020).

Availability of modelling data

R_0 has been calculated for a number of species of disease vectors including *Stomoxys calcitrans*, *Culicoides nubeculosus* and three mosquito species: *Aedes aegypti*, *Anopheles stephensi* and *Culex quinquefasciatus* (Gubbins, 2019). A report by EFSA (2020) considered a number of important transmission parameters, simulated within-herd prevalence based on the bites of *S. calcitrans* and calculated the probabilities of LSD introduction via imported cattle. An earlier study focusing on dairy herds in Israel calculated estimates of R_0 based on both direct (0.36) and indirect (15.7) transmission (Magori-Cohen *et al.*, 2012).

Avian Infectious Bronchitis

Overview

Avian infectious bronchitis (IB) is a disease primarily of poultry caused by the Gammacoronavirus infectious bronchitis virus (IBV). IBV initially infects the respiratory tract, leading to clinical signs such as coughing, sneezing, tracheal and nasal discharge, and dyspnoea. IBV can become disseminated to other tissues, including the renal and reproductive systems (Zhao *et al.*, 2023). Dissemination to other tissues is thought to be mediated by monocyte infection (Reddy *et al.*, 2016) and/or via macrophages and the lymphatic system (Amarasinghe *et al.*, 2017). Nephropathogenic strains of IBV have attracted more attention in recent years due to their increased virulence in young birds (Feng *et al.*, 2012). IBV infections in young birds typically occur between 1 to 30 days of age and can lead to reproductive tract defects, with subsequent low egg-laying production and poor egg quality (Zhao *et al.*, 2023). Although tissue tropism varies between IBV strains, there are also strains which simultaneously infect multiple systems, and those that infect primarily the respiratory tract can lead to secondary bacterial infections (Zhao *et al.*, 2023).

Prevalence

Infectious bronchitis virus is globally distributed but different strains predominate in different countries and regions (Zhao *et al.*, 2023). Seroprevalence studies from Spain indicate that IBV was found in 38% of broilers, 44% of layers and 43% of broiler breeders (Cortés *et al.*, 2022). A separate report from hobby poultry in Flanders, Belgium, found IBV in 75.6% of samples (Haesendonck *et al.*, 2014). IBV was found in 20% of flocks sampled in Greece (Andreopoulou *et al.*, 2019). Earlier data from France indicated that IBV was present in 61% of flocks sampled (Auvigne *et al.*, 2013).

Risk factors

Despite vaccination for IBR being widespread in commercial poultry flocks, vaccines are not 100% effective due to the high mutation and recombination rate of IBV (Zhao *et al.*, 2023). As such, even within vaccinated flocks there is often circulation and risk of introduction, particularly via transport of live poultry, sharing of personnel between facilities and contamination of fomites and transport vehicles (Franzo *et al.*, 2020).

Transmission

Infectious bronchitis virus is highly infectious, with infected chicks being responsible for the majority of environmental contamination. Transmission of IBV is primarily through respiratory secretions and faeces, with additional spread through environmental contamination particularly of equipment, egg-packing material and by farm visitors and personnel (Ramakrishnan and Kappala, 2019).

Control measures and biosecurity

Vaccination, despite its lack of 100% protection, remains a mainstay of IBV control (Franzo *et al.*, 2020; Khera *et al.*, 2022; Zhao *et al.*, 2023). In many countries. this is done at 1 day old, with chicks immediately vaccinated (Khera *et al.*, 2022). Beyond this, proper biosecurity and management control are essential. Measures include maintaining proper bird density, good air flow and quality, and strict biosecurity measures (Franzo *et al.*, 2020; Khera *et al.*, 2022). Other measures include primary and secondary containment including limited site access, providing proper boots and uniforms for staff, farm equipment and litter sanitation, foot and wheel baths for incoming vehicles and personnel which should be placed at all entrances, the enforcement of strict cleaning, avoiding repeated use of the same disinfectants in 6–12 month periods, and proper sanitization of housing facilities, with 10–14 day-downtime periods between housing periods (Bhuiyan *et al.*, 2021). The widespread integration within the poultry industry can facilitate this (Franzo *et al.*, 2020). In line with this, it is also of critical importance to maintain an appropriate distance between poultry farms to prevent IBV infections spreading (Bhuiyan *et al.*, 2021).

Availability of modelling data

The R_0 for IBV in chickens has been estimated at 19.95, with vaccination leading to a significant reduction in transmission (R = 0.69) (de Wit *et al.*, 1998). Some transmission parameters have also been estimated for a novel coronavirus in turkeys (Brown *et al.*, 2019).

Porcine Epidemic Diarrhoea Virus

Overview

Porcine epidemic diarrhoea (PED) is a highly contagious intestinal disease caused by the porcine epidemic diarrhoea virus (PEDV) (Zhang *et al.*, 2022), an alphacoronavirus infecting domestic swine (Liu and Wang, 2021). PED mainly affects piglets, with disease aetiology generally more severe than in adults, with clinical signs such as vomiting, watery diarrhoea, dehydration, anorexia and weight loss being typical (Zhang *et al.*, 2022). Although the overall mortality is quite low, typically less than 5%, in piglets up to 2 weeks of age it can be as high as 50% (Lee, 2015; Zhang *et al.*, 2022). Economic losses are estimated at €127–650 per sow (Renken *et al.*, 2021).

Prevalence

Porcine epidemic diarrhoea virus remains endemic in Asia but outbreaks in the preceding three decades in Europe were rare (Karte *et al.*, 2020). However, since the emergence of a highly virulent strain in the United States in 2013, with explosive outbreaks affecting all ages of animals occurring in 31 of the 50 states, the disease has re-emerged in Europe (Hanke *et al.*, 2017; Antas and Woźniakowski, 2019; Karte *et al.*, 2020). PED re-emerged in Germany in 2014 (Hanke *et al.*, 2017; Karte *et al.*, 2020) and has subsequently spread to Italy (Ferrara *et al.*, 2023), France (Gallien *et al.*, 2018), Belgium (Theuns *et al.*, 2015), Austria (Steinrigl *et al.*, 2015) and Spain (Hanke *et al.*, 2017).

Risk factors

Based on the current epidemiological knowledge of PEDV spread, it appears that the most significant risk factor for the introduction of PEDV to a farm is via the movement of animals and the contamination of transport vehicles (Trostle *et al.*, 2022). This risk has been particularly associated with the purchase of replacement gilts (VanderWaal *et al.*, 2018). As with many other diseases, contamination and transmission via fomites (Lowe *et al.*, 2014) also necessitate strict biosecurity and appropriate disinfection procedures for equipment and farm personnel. An analysis of risk factors during the 2013–2014 PEDV outbreak in Japan identified previous occurrence of PRRS within the past 5 years, use of a common compost station, and use of a pig excrement disposal service as risk factors for the introduction of PEDV to a farm (Toyomaki *et al.*, 2018).

Transmission

Porcine epidemic diarrhoea virus can be transmitted directly via the faecal–oral and faecal–nasal routes (Jung and Saif, 2015), vertically through sow milk (Sun *et al.*, 2012), sexual intercourse or artificial insemination (Gallien *et al.*, 2018) and via aerosol transmission (Zhang *et al.*, 2022). In addition, contaminated vehicles, farm equipment and farm personnel constitute indirect routes of transmission (Lowe *et al.*, 2014; Bowman *et al.*, 2015). Contaminated feed was implicated as a potential transmission route in the 2014 PEDV outbreak in Canada; although experimental data showed low shedding, the route could nonetheless not be ruled out (Pasick *et al.*, 2014). Contaminated feed has also been posited as having played a potential role in an outbreak in a breeding herd in Mexico (Garrido-Mantilla *et al.*, 2022).

Control measures and biosecurity

Given the critical role played by animal transport practices and external personnel in the dispersal of PEDV, the most effective biosecurity measures centre on tackling this aspect of external biosecurity for a farm. Ideal measures to combat the introduction of PEDV include the disinfection of vehicle trailers and transport spaces prior to each delivery of pigs; drivers remaining within the vehicle at all times during the delivery, minimizing risk of becoming contaminated at the farm; enforcement of a shower-in policy or bench entry; all staff and visitors being required to wear specific barn clothing (Perri *et al.*, 2018).

Availability of modelling data

Analysis of epidemiological data from the United States estimated a number of transmission rates for both local (1.2), and via movements (0.014), the indirect transmission rate (0.53), individual shedding rate (0.27) and the reproductive rate (0.65), and assessed the efficacy of various biosecurity measures and identified areas of particular risk (Galvis *et al.*, 2022). Data modelled from the US also evaluated the efficacy of biosecurity measures to prevent indirect transmission of PEDV (Kim *et al.*, 2017).

Analysis of epidemiological data from the 2014 PEDV outbreak in Ontario, Canada, found that the R_0 (herein defined as farm-to-farm transmissions) was estimated at 1.87 for the base case, with estimates of 2.21–2.8 if 25% or 50% underreporting of cases is assumed, and up to 4.92 if a 13 day generation time is assumed (Greer *et al.*, 2017). The infectious period for PEDV has been estimated at 54.4 days, and R_0 in the same study was estimated to be 5.39 for the 2013–2014 outbreak in Japan (Makita and Yamamoto, 2019). The indirect transmission rate has been calculated at 1.22×10^{-12} for PEDV based on Japanese empirical data and the effect of various control measures modelled for a farrow-finish farm (Murai *et al.*, 2018), with a further analysis based on the Japanese 2013–2014 outbreak detailing risk factors associated with farm-level infection with PEDV (Toyomaki *et al.*, 2018) and the role of epidemiological factors in PEDV spread (Sasaki *et al.*, 2016).

A further study examined the role of transport practices in the spread of PEDV in Italy (Boniotti *et al.*, 2018). The survival of PEDV in on-farm earthen manure storage has been explored, finding survival of up to 9 months, although this varied by layer within the manure storage due to factors such as UV penetration and other environmental effects (Tun *et al.*, 2016). Finally, thermal inactivation of PEDV has been explored, suggesting that optimal temperatures of >76°C were necessary for full inactivation (Zentkovich *et al.*, 2016).

Brucellosis

Overview

Brucellosis is a disease of significant veterinary and public health concern (Khurana *et al.*, 2021; Katsiolis *et al.*, 2022), in addition to constituting an important zoonosis (Katsiolis *et al.*, 2022). The aetiological agent affecting cattle is *Brucella abortus*, with *B. melitensis* in small ruminants and *B. suis* in swine (Jamil *et al.*, 2022), although the latter two can also cause spillover infection in cattle (Khurana *et al.*, 2021; Jamil *et al.*, 2022). Clinical signs can be varied, although typically the defining clinical sign, particularly for *B. abortus* in cattle, is spontaneous abortion, in addition to orchitis and epididymitis in male animals (Khurana *et al.*, 2021).

Prevalence in Spain, Croatia, Greece, Italy and Portugal

Brucellosis has been successfully eradicated within much of the European Union in farm animals (Jamil *et al.*, 2022). Eradication is near completion in Spain and Croatia, but persists as a serious veterinary and public health concern in Greece, Italy and Portugal, albeit with declining incidence rates (Wareth *et al.*, 2019; EFSA, 2021). Approximately 0.38% of bovine herds (603/157,000) and 0.22% of small ruminant herds (349/160,000) still test positive for *Brucella* spp. in endemic regions of the EU (EFSA, 2021).

Risk factors

Mixed farming of bovine and small ruminants has been identified as a significant risk factor for brucellosis, particularly for *B. melitsensis*, which primarily affects small ruminants but with cattle as a potential spillover host (Khurana *et al.*, 2021). Lack of awareness surrounding brucellosis and its risk constitutes a further factor both for its spread within herds and for zoonotic potential (Machavarapu *et al.*, 2019). Immunization with attenuated vaccine strains has proved effective at controlling brucellosis in many endemic regions, but this must be combined with effective diagnostic surveillance, mass vaccination of large numbers of the population to achieve an effective level of immunity, and consistent culling of *Brucella*-positive animals (Khurana *et al.*, 2021). This is the current method of control employed in endemic countries within the EU, such as Greece (Katsiolis *et al.*, 2022).

Transmission

Brucella spp. can be transmitted in livestock both horizontally and vertically (Meltzer *et al.*, 2010), with the uterus of infected pregnant animals showing particularly high levels of the bacterium (Khurana *et al.*, 2021). Aborted fetuses and placenta of infected animals also constitute a significant source of contamination of the environment with *Brucella* spp. Shedding of bacteria in the milk of infected animals can transmit the bacterium to newborns and young animals. *Brucella* spp. are known to persist in the environment for months, particularly in cold and moist weather, raising the risk of environmental transmission via contaminated pasture or housing (Khurana *et al.*, 2021). The sharing of contaminated water sources is a significant cause of *Brucella* spread (Tukana and Gummow, 2017). Finally, infected bulls are capable of transmitting the bacterium venereally, and artificial insemination with semen from an infected bull can transmit the bacterium to the female (Khurana *et al.*, 2021).

Control measures and biosecurity

Control strategies for brucellosis include effective disease surveillance, the prevention of transmission (Khurana *et al.*, 2021), which is of particular importance regarding aborted fetuses in which very high numbers of bacteria are present, and controlling the reservoir of infection by culling (Durrani *et al.*, 2020; Khurana *et al.*, 2021).

Availability of modelling data

Blackburn *et al.* (2019) produced a detailed review covering the calculation of R_0 for pathogens with environmental transmission, among which *Brucella* spp. are considered. The dynamics and transmission of dairy cattle brucellosis have been modelled, and R_0 calculated, for Jilin Province, China (Nie *et al*., 2014), and the effect of transport on brucellosis transmission in China (Liang *et al*., 2020). Lou *et al*. (2016) modelled the transmission of brucellosis in sheep in Inner Mongolia Province, China. An SEIB model of brucellosis transmission in dairy cows with an immunological threshold has been constructed (Zhang *et al*., 2019).

Abagna *et al*. (2022) modelled bovine brucellosis and the effect of various control measures, whereas Nyerere *et al*. (2020) modelled the impact of various weather conditions and seasonal events on brucellosis transmission. Tumwiine and Robert (2017) modelled the treatment of brucellosis in the cattle population. Liang and Fang (2022) modelled transmission of brucellosis with an incubation period. Finally, Wang *et al*. (2015) considered the *in vitro* effects of six disinfectants on *B. melitensis*, and their effects on a variety of conditions including presence of moisture, dirt, exposure time and temperature.

References

Abagna, S., Seidu, B. and Bornaa, C. (2022) A mathematical model of the transmission dynamics and control of bovine brucellosis in cattle. *Abstract and Applied Analysis* 1–10.

Abdo-Salem, S., Gerbier, G., Bonnet, P., Al-Qadasi, M., Tran, A. *et al*. (2006) Descriptive and spatial epidemiology of Rift Valley fever outbreak in Yemen 2000–2001. *Annals of the New York Academy of Sciences* 1081(1), 240–242.

Abdrakhmanov, S.K., Beisembayev, K.K., Sultanov, A.A., Mukhanbetkaliyev, Y.Y., Kadyrov, A.S. *et al*. (2021) Modelling bluetongue risk in Kazakhstan. *Parasites & Vectors* 14(1), 491.

Abutarbush, S.M. (2014) Efficacy of vaccination against lumpy skin disease in Jordanian cattle. *Veterinary Record* 175(12), 302.

Adkin, A., England, T., Hall, S., Coburn, H., Marooney, C.J. *et al*. (2008) Estimating the risk of exposure of British livestock to foot-and-mouth disease associated with the importation of ship and aircraft waste. *Veterinary Record* 163(8), 235–240.

Agianniotaki, E.I., Tasioudi, K.E., Chaintoutis, S.C., Iliadou, P., Mangana-Vougiouka, O. *et al*. (2017) Lumpy skin disease outbreaks in Greece during 2015–16, implementation of emergency immunization and genetic differentiation between field isolates and vaccine virus strains. *Veterinary Microbiology* 201, 78–84.

Alarcón, L.V., Allepuz, A. and Mateu, E. (2021) Biosecurity in pig farms: A review. *Porcine Health Management* 7(1), 5.

Albayrak, H. and Alkan, F. (2009) PPR virus infection on sheep in Blacksea Region of Turkey: Epidemiology and diagnosis by RT-PCR and virus isolation. *Veterinary Research Communications* 33(3), 241–249.

Aleksandr, K., Olga, B., David, W.B., Pavel, P., Yana, P. *et al*. (2020) Non-vector-borne transmission of lumpy skin disease virus. *Scientific Reports* 10(1), 7436.

Alexandersen, S. and Donaldson, A.I. (2002) Further studies to quantify the dose of natural aerosols of foot-and-mouth disease virus for pigs. *Epidemiology and Infection* 128(2), 313–323.

Alexandersen, S., Brotherhood, I. and Donaldson, A.I. (2002) Natural aerosol transmission of foot-and-mouth disease virus to pigs: Minimal infectious dose for strain O1 Lausanne. *Epidemiology and Infection* 128(2), 301–312.

Alexandersen, S., Quan, M., Murphy, C., Knight, J. and Zhang, Z. (2003a) Studies of quantitative parameters of virus excretion and transmission in pigs and cattle experimentally infected with foot-and-mouth disease virus. *Journal of Comparative Pathology* 129(4), 268–282.

Alexandersen, S., Zhang, Z., Donaldson, A.I. and Garland, A.J.M. (2003b) The pathogenesis and diagnosis of foot-and-mouth disease. *Journal of Comparative Pathology* 129(1), 1–36.

Altan, E., Parida, S., Mahapatra, M., Turan, N. and Yilmaz, H. (2019) Molecular characterization of peste des petits ruminants viruses in the Marmara Region of Turkey. *Transboundary and Emerging Diseases* 66(2), 865–872.

Amarasinghe, A., Abdul-Cader, M.S., Nazir, S., Silva Senapathi, U., Meer, F. *et al.* (2017) Infectious bronchitis corona virus establishes productive infection in avian macrophages interfering with selected antimicrobial functions. *PLoS One* 12(8), e0181801.

Andreopoulou, M., Franzo, G., Tucciarone, C.M., Prentza, Z., Koutoulis, K.C. *et al.* (2019) Molecular epidemiology of infectious bronchitis virus and avian metapneumovirus in Greece. *Poultry Science* 98(11), 5374–5384.

Annandale, C.H., Holm, D.E., Ebersohn, K. and Venter, E.H. (2014) Seminal transmission of lumpy skin disease virus in heifers. *Transboundary and Emerging Diseases* 61(5), 443–448.

Antas, M. and Woźniakowski, G. (2019) Current status of porcine epidemic diarrhoea (PED) in European pigs. *Journal of Veterinary Research* 63(4), 465–470.

Apolloni, A., Bataille, A., Coste, C., Diallo, A., Hammami, P. *et al.* (2016) PPR vaccination strategy with a special reference to West Africa. Available at: www.researchgate.net/publication/313609862_PPR_vaccination_strategy_with_a_special_reference_to_West_Africa (accessed 11 December 2024).

Assefa, A., Tibebu, A., Bihon, A. and Yimana, M. (2021) Global ecological niche modelling of current and future distribution of peste des petits ruminants virus (pprv) with an ensemble modelling algorithm. *Transboundary and Emerging Diseases* 68(6), 3601–3610.

Auvigne, V., Gibaud, S., Léger, L., Malher, X., Currie, R. *et al.* (2013) A longitudinal study of the incidence of avian infectious bronchitis in France using strain-specific haemagglutination inhibition tests and cluster analysis. *RNA* 4, 11.

Balkhy, H.H. and Memish, Z.A. (2003) Rift Valley fever: An uninvited zoonosis in the Arabian Peninsula. *International Journal of Antimicrobial Agents* 21(2), 153–157.

Bandou, R., Hirose, R., Nakaya, T., Miyazaki, H., Watanabe, N. *et al.* (2022) Higher viral stability and ethanol resistance of avian influenza A(H5N1) virus on human skin. *Emerging Infectious Diseases* 28(3), 639–649.

Barlow, D. (1972) The effects of various protecting agents on the inactivation of foot-and-mouth disease virus in aerosols and during freeze-drying. *Journal of General Virology* 17(3), 281–288.

Barlow, D.F. and Donaldson, A.I. (1973) Comparison of the aerosol stabilities of foot-and-mouth disease virus suspended in cell culture fluid or natural fluids. *Journal of General Virology* 20(3), 311–318.

Baron, M.D., Diallo, A., Lancelot, R. and Libeau, G. (2016) Peste des petits ruminants virus. In: Kielian, M., Maramorosch, K. and Mettenleiter, T.C. (eds) *Advances in Virus Research*, Vol. 95. Academic Press, New York, pp. 1–42.

Barongo, M.B., Bishop, R.P., Fèvre, E.M., Knobel, D.L. and Ssematimba, A. (2016) A mathematical model that simulates control options for African swine fever virus (ASFV). *PLoS One* 11(7), e0158658.

Bates, T.W., Thurmond, M.C. and Carpenter, T.E. (2003) Description of an epidemic simulation model for use in evaluating strategies to control an outbreak of foot-and-mouth disease. *American Journal of Veterinary Research* 64(2), 195–204.

Baylis, M., O'Connell, L. and Mellor, P.S. (2008) Rates of bluetongue virus transmission between *Culicoides sonorensis* and sheep. *Medical and Veterinary Entomology* 22(3), 228–237.

Beato, M.S., D'Errico, F., Iscaro, C., Petrini, S., Giammarioli, M. *et al.* (2022) Disinfectants against African swine fever: An updated review. *Viruses* 14(7), 1384.

Bech-Nielsen, S., Perez Bonilla, Q. and Sanchez-Vizcaino, J.M. (1993) Benefit–cost analysis of the current African swine fever eradication program in Spain and of an accelerated program. *Preventive Veterinary Medicine* 17(3), 235–249.

Ben Chehida, F., Ayari-Fakhfakh, E., Caufour, P., Amdouni, J., Nasr, J. *et al.* (2018) Sheep pox in Tunisia: Current status and perspectives. *Transboundary and Emerging Diseases* 65(1), 50–63.

Ben-Gera, J., Klement, E., Khinich, E., Stram, Y. and Shpigel, N.Y. (2015) Comparison of the efficacy of Neethling lumpy skin disease virus and x10rm65 sheep-pox live attenuated vaccines for the prevention of lumpy skin disease – the results of a randomized controlled field study. *Vaccine* 33(38), 4837–4842.

Bergmann, H., Schulz, K., Conraths, F.J. and Sauter-Louis, C. (2021) A review of environmental risk factors for African swine fever in European wild boar. *Animals* 11(9), 2692.

Bergmann, H., Dups-Bergmann, J., Schulz, K., Probst, C., Zani, L. *et al.* (2022) Identification of risk factors for African swine fever: A systematic review. *Viruses* 14(10), 2107.

Bhuiyan, M.S.A., Amin, Z., Rodrigues, K.F., Saallah, S., Shaarani, S.M. *et al.* (2021) Infectious bronchitis virus (Gammacoronavirus) in poultry farming: Vaccination. *Veterinary Sciences* 8(11), 273.

Bird, B.H., Ksiazek, T.G., Nichol, S.T. and MacLachlan, N.J. (2009) Rift Valley fever virus. *Journal of the American Veterinary Medical Association* 234(7), 883–893.

Björnham, O., Sigg, R. and Burman, J. (2020) Multilevel model for airborne transmission of foot-and-mouth disease applied to Swedish livestock. *PLoS One* 15(5), e0232489.
Blackburn, J.K., Ganz, H.H., Ponciano, J.M., Turner, W.C., Ryan, S.J. *et al.* (2019) Modeling R0 for pathogens with environmental transmission: Animal movements, pathogen populations, and local infectious zones. *International Journal of Environmental Research and Public Health* 16(6), 954.
Blagodatski, A., Trutneva, K., Glazova, O., Mityaeva, O., Shevkova, L. *et al.* (2021) Avian influenza in wild birds and poultry: Dissemination pathways, monitoring methods, and virus ecology. *Pathogens* 10(5), 630.
Blome, S., Gabriel, C. and Beer, M. (2013) Pathogenesis of African swine fever in domestic pigs and European wild boar. *Virus Research* 173(1), 122–130.
Blome, S., Staubach, C., Henke, J., Carlson, J. and Beer, M. (2017) Classical swine fever – an updated review. *Viruses* 9(4), 86.
Boender, G.J., Hengel, R., Roermund, H.J.W. and Hagenaars, T.J. (2014) The influence of between-farm distance and farm size on the spread of classical swine fever during the 1997–1998 epidemic in the Netherlands. *PLoS One* 9(4), e95278.
Boinas, F.S., Wilson, A.J., Hutchings, G.H., Martins, C. and Dixon, L.J. (2011) The persistence of African swine fever virus in field-infected *Ornithodoros erraticus* during the ASF endemic period in Portugal. *PLoS One* 6(5), e20383.
Boniotti, M.B., Papetti, A., Bertasio, C., Giacomini, E., Lazzaro, M. *et al.* (2018) Porcine epidemic diarrhoea virus in Italy: Disease spread and the role of transportation. *Transboundary and Emerging Diseases* 65(6), 1935–1942.
Bowden, T.R., Babiuk, S.L., Parkyn, G.R., Copps, J.S. and Boyle, D.B. (2008) Capripoxvirus tissue tropism and shedding: A quantitative study in experimentally infected sheep and goats. *Virology* 371(2), 380–393.
Bowman, A.S., Krogwold, R.A., Price, T., Davis, M. and Moeller, S.J. (2015) Investigating the introduction of porcine epidemic diarrhea virus into an Ohio swine operation. *BMC Veterinary Research* 11(1), 38.
Bradhurst, R., Garner, G., Hóvári, M., Puente, M., Mintiens, K. *et al.* (2022) Development of a transboundary model of livestock disease in Europe. *Transboundary and Emerging Diseases* 69(4), 1963–1982.
Bravo de Rueda, C., de Jong, M.C.M., Eblé, P.L. and Dekker, A. (2015) Quantification of transmission of foot-and-mouth disease virus caused by an environment contaminated with secretions and excretions from infected calves. *Veterinary Research* 46(1), 43.
Brown, E., Nelson, N., Gubbins, S. and Colenutt, C. (2022) Airborne transmission of foot-and-mouth disease virus: A review of past and present perspectives. *Viruses* 14(5), 1009.
Brown, P.A., Courtillon, C., Weerts, E.A.W.S., Andraud, M., Allée, C. *et al.* (2019) Transmission kinetics and histopathology induced by European Turkey coronavirus during experimental infection of specific pathogen free Turkeys. *Transboundary and Emerging Diseases* 66(1), 234–242.
Brown, V.R., Miller, R.S., McKee, S.C., Ernst, K.H., Didero, N.M. *et al.* (2021) Risks of introduction and economic consequences associated with African swine fever, classical swine fever and foot-and-mouth disease: A review of the literature. *Transboundary Emerging Disease* 68, 1910–1965. DOI: 10.1111/tbed.13919.
Bui, C., Bethmont, A., Chughtai, A.A., Gardner, L., Sarkar, S. *et al.* (2016) A systematic review of the comparative epidemiology of avian and human influenza A H5N1 and H7N9 – lessons and unanswered questions. *Transboundary and Emerging Diseases* 63(6), 602–620.
Carpenter, T.E., Thurmond, M.C. and Bates, T.W. (2004) A simulation model of intraherd transmission of foot and mouth disease with reference to disease spread before and after clinical diagnosis. *Journal of Veterinary Diagnostic Investigation* 16(1), 11–16.
Cecilia, H., Métras, R., Fall, A.G., Lo, M.M., Lancelot, R. *et al.* (2020) It's risky to wander in September: Modelling the epidemic potential of Rift Valley fever in a sahelian setting. *Epidemics* 33, 100409.
Charron, M.V., Kluiters, G., Langlais, M., Seegers, H., Baylis, M. *et al.* (2013) Seasonal and spatial heterogeneities in host and vector abundances impact the spatiotemporal spread of bluetongue. *Veterinary Research* 44(1), 44.
Chatziprodromidou, I.P., Arvanitidou, M., Guitian, J., Apostolou, T., Vantarakis, G. *et al.* (2018) Global avian influenza outbreaks 2010–2016: A systematic review of their distribution, avian species and virus subtype. *Systematic Reviews* 7(1), 17.
Chenais, E., Depner, K., Guberti, V., Dietze, K., Viltrop, A. *et al.* (2019) Epidemiological considerations on African swine fever in Europe 2014–2018. *Porcine Health Management* 5, 6.

Chevalier, V., Pépin, M., Plée, L. and Lancelot, R. (2010) Rift Valley fever – a threat for Europe. *Euro Surveillance* 15(10), 19506.

Chis Ster, I., Dodd, P.J. and Ferguson, N.M. (2012) Within-farm transmission dynamics of foot and mouth disease as revealed by the 2001 epidemic in Great Britain. *Epidemics* 4(3), 158–169.

Chitnis, N., Hyman, J.M. and Manore, C.A. (2013) Modelling vertical transmission in vector-borne diseases with applications to Rift Valley fever. *Journal of Biological Dynamics* 7(1), 11–40.

Chmitelin, I. and Moutou, F. (2002) Foot and mouth disease: Lessons to be learned from the experience of France. *Revue Scientifique et Technique (International Office of Epizootics)* 21(3), 731–737.

Clements, A.C., Pfeiffer, D.U. and Martin, V. (2006) Application of knowledge-driven spatial modelling approaches and uncertainty management to a study of Rift Valley fever in Africa. *International Journal of Health Geographics* 5, 57.

Coffman, M.S., Sanderson, M.W., Dodd, C.C., Arzt, J. and Renter, D.G. (2021) Estimation of foot-and-mouth disease windborne transmission risk from USA beef feedlots. *Preventive Veterinary Medicine* 195, 105453.

Colenutt, C., Gonzales, J.L., Paton, D.J., Gloster, J., Nelson, N. *et al.* (2016) Aerosol transmission of foot-and-mouth disease virus Asia-1 under experimental conditions. *Veterinary Microbiology* 189, 39–45.

Colenutt, C., Brown, E., Nelson, N., Paton, D.J., Eblé, P. *et al.* (2020) Quantifying the transmission of foot-and-mouth disease virus in cattle via a contaminated environment. *mBio* 11(4), e00381–20.

Coronado, L., Perera, C.L., Rios, L., Frías, M.T. and Pérez, L.J. (2021) A critical review about different vaccines against classical swine fever virus and their repercussions in endemic regions. *Vaccines (Basel)* 9(2), 154. DOI: 10.3390/vaccines9020154.

Cortés, V., Sevilla-Navarro, S., García, C., Marín, C. and Catalá-Gregori, P. (2022) Seroprevalence and prevalence of infectious bronchitis virus in broilers, laying hens and broiler breeders in Spain. *Poultry Science* 101(5), 101760.

Costard, S., Mur, L., Lubroth, J., Sanchez-Vizcaino, J.M. and Pfeiffer, D.U. (2013) Epidemiology of African swine fever virus. *Virus Research* 173(1), 191–197.

Couacy-Hymann, E., Bodjo, S.C., Danho, T., Koffi, M.Y., Libeau, G. *et al.* (2007a) Early detection of viral excretion from experimentally infected goats with peste-des-petits ruminants virus. *Preventive Veterinary Medicine* 78(1), 85–88.

Couacy-Hymann, E., Bodjo, C., Danho, T., Libeau, G. and Diallo, A. (2007b) Evaluation of the virulence of some strains of peste-des-petits-ruminants virus (PPRV) in experimentally infected West African dwarf goats. *Veterinary Journal* 173(1), 178–183.

Courtejoie, N., Zanella, G. and Durand, B. (2018) Bluetongue transmission and control in Europe: A systematic review of compartmental mathematical models. *Preventive Veterinary Medicine* 156, 113–125.

Courtejoie, N., Cauchemez, S., Zanella, G. and Durand, B. (2019) A network-based approach to modelling bluetongue spread in France. *Preventive Veterinary Medicine* 170, 104744.

Cwynar, P., Stojkov, J. and Wlazlak, K. (2019) African swine fever status in Europe. *Viruses* 11(4), 310.

De Benedictis, P., Beato, M.S. and Capua, I. (2007) Inactivation of avian influenza viruses by chemical agents and physical conditions: A review. *Zoonoses and Public Health* 54(2), 51–68.

de Carvalho Ferreira, H.C., Backer, J.A., Weesendorp, E., Klinkenberg, D., Stegeman, J.A. *et al.* (2013) Transmission rate of African swine fever virus under experimental conditions. *Veterinary Microbiology* 165(3–4), 296–304.

De Clercq, K., Vandaele, L., Vanbinst, T., Riou, M., Deblauwe, I. *et al.* (2021) Transmission of bluetongue virus serotype 8 by artificial insemination with frozen-thawed semen from naturally infected bulls. *Viruses* 13(4), 652.

DEFRA (2023) Preliminary outbreak assessment #2: Sheep and goat pox in Europe. Available at: https://assets.publishing.service.gov.uk/government/uploads/system/uploads/attachment_data/file/1162857/6_June_2023_Sheep_and_goat_pox_in_Europe.pdf (accessed 11 December 2024).

de la Rocque, S., Tran, A.L., Etter, E., Vial, L. and Hendrickx, G. (2007) Environmental changes, disease ecology and geographic information system-based tools for risk assessment. *Veterinaria Italiana* 43(3), 381–391.

de la Torre, A., Bosch, J., Sánchez-Vizcaíno, J.M., Ito, S., Muñoz, C. *et al.* (2022) African swine fever survey in a European context. *Pathogens* 11(2), 137.

de Wit, J.J., de Jong, M.C., Pijpers, A. and Verheijden, J.H. (1998) Transmission of infectious bronchitis virus within vaccinated and unvaccinated groups of chickens. *Avian Pathology* 27(5), 464–471.

Dhollander, S., Belsham, G.J., Lange, M., Willgert, K., Alexandrov, T. *et al.* (2016) Assessing the potential spread and maintenance of foot-and-mouth disease virus infection in wild ungulates: General principles and application to a specific scenario in Thrace. *Transboundary and Emerging Diseases* 63(2), 165–174.

Dillon, M.B. (2011) Skin as a potential source of infectious foot and mouth disease aerosols. *Proceedings. Biological Sciences* 278(1713), 1761–1769.

Dion, E., Van Schalkwyk, L. and Lambin, E.F. (2011) The landscape epidemiology of foot-and-mouth disease in South Africa: A spatially explicit multi-agent simulation. *Ecological Modelling* 222(13), 2059–2072.

Diop, M., Sarr, J. and Libeau, G. (2005) Evaluation of novel diagnostic tools for peste des petits ruminants virus in naturally infected goat herds. *Epidemiology and Infection* 133(4), 711–717.

Dixon, L.K., Stahl, K., Jori, F., Vial, L. and Pfeiffer, D.U. (2020) African swine fever epidemiology and control. *Annual Review of Animal Biosciences* 8(1), 221–246.

Donaldson, A.I., Gloster, J., Harvey, L.D. and Deans, D.H. (1982) Use of prediction models to forecast and analyse airborne spread during the foot-and-mouth disease outbreaks in Brittany, Jersey and the Isle of Wight in 1981. *Veterinary Record* 110(3), 53–57.

Doran, R.J. and Laffan, S.W. (2005) Simulating the spatial dynamics of foot and mouth disease outbreaks in feral pigs and livestock in Queensland, Australia, using a susceptible-infected-recovered cellular automata model. *Preventive Veterinary Medicine* 70(1–2), 133–152.

Dorigatti, I., Mulatti, P., Rosà, R., Pugliese, A. and Busani, L. (2010) Modelling the spatial spread of H7N1 avian influenza virus among poultry farms in Italy. *Epidemics* 2, 29–35. DOI: 10.1016/j.epidem.2010.01.002.

Durand, B., Davila, S., Cariolet, R., Mesplède, A. and Le Potier, M.F. (2009) Comparison of viraemia- and clinical-based estimates of within- and between-pen transmission of classical swine fever virus from three transmission experiments. *Veterinary Microbiology* 135(3–4), 196–204.

Durand, B., Lo Modou, M., Tran, A., Ba, A., Sow, F. *et al.* (2020) Rift Valley fever in northern Senegal: A modelling approach to analyse the processes underlying virus circulation recurrence. *PLoS Neglected Tropical Diseases* 14(6), e0008009.

Durrani, A.Z., Usman, M., Kazmi, Z. and Husnain, M. (2020) Evaluation of therapeutic trials in bovines. Available at: www.uv.mx/personal/avillagomez/files/2020/01/2020-Brucella-Book-chapter.pdf (accessed 11 December 2024).

Eagles, D., Melville, L., Weir, R., Davis, S., Bellis, G. *et al.* (2014) Long-distanceaerial dispersal modelling of *Culicoides* biting midges: Case studies of incursions into Australia. *BMC Veterinary Research* 10(1), 135.

Eblé, P., De Koeijer, A., Bouma, A., Stegeman, A. and Dekker, A. (2006) Quantification of within-and between-pen transmission of foot-and-mouth disease virus in pigs. *Veterinary Research* 37(5), 647–654.

Eblé, P.L., Hagenaars, T.J., Weesendorp, E., Quak, S., Moonen-Leusen, H.W. *et al.* (2019) Transmission of African swine fever virus via carrier (survivor) pigs does occur. *Veterinary Microbiology* 237, 108345.

EFSA Panel on Animal Health and Welfare (AHAW) (2012a) Scientific opinion on swine vesicular disease and vesicular stomatitis. *EFSA Journal* 10(4), 2631.

EFSA Panel on Animal Health and Welfare (AHAW) (2012b) Scientific opinion on foot-and-mouth disease in Thrace. *EFSA Journal* 10(4), 2635.

EFSA Panel on Animal Health and Welfare (AHAW), More, S., Bicout, D., Bøtner, A., Butterworth, A. *et al.* (2017a) Assessment of listing and categorisation of animal diseases within the framework of the animal health law (Regulation (EU) No 2016/429): Bluetongue. *EFSA Journal* 15(8), e04957.

EFSA Panel on Animal Health and Welfare (AHAW) (2017b) Bluetongue: Control, surveillance and safe movement of animals. *EFSA Journal* 15(3), e04698.

EFSA Panel on Animal Health and Welfare (AHAW), Nielsen, S.S., Alvarez, J., Bicout, D.J., Calistri, P. *et al.* (2021) Assessment of the control measures of the category A diseases of animal health law: Classical swine fever. *EFSA Journal* 19(7), e06707. DOI: 10.2903/j.efsa.2021.6707.

Endris, R.G. and Hess, W.R. (1992) Experimental transmission of African swine fever virus by the soft tick *Ornithodoros* (Pavlovskyella) *marocanus* (Acari: Ixodoidea: Argasidae). *Journal of Medical Entomology* 29(4), 652–656.

Esser, H.J., Liefting, Y., Ibáñez-Justicia, A., van der Jeugd, H., van Turnhout, C.A.M. *et al.* (2020) Spatial risk analysis for the introduction and circulation of six arboviruses in The Netherlands. *Parasites & Vectors* 13(1), 464.

European Food Safety Authority (EFSA), Calistri, P., De Clercq, K., Gubbins, S., Klement, E. *et al.* (2020) Lumpy skin disease epidemiological report IV: Data collection and analysis. *EFSA Journal* 18(2), e06010. DOI: 10.2903/j.efsa.2020.6010.

European Food Safety Authority (EFSA) (2021) The European Union One health 2019 zoonoses report. *EFSA Journal* 19(2), e06406.

European Food Safety Authority (EFSA), Baños, J.V., Boklund, A., Gogin, A., Gortázar, C. *et al.* (2022) Scientific report on the epidemiological analyses of African swine fever in the European Union. *EFSA Journal* 20(5), 106. DOI: 10.2903/j.efsa.2022.7290.

European Food Safety Authority (EFSA), Adlhoch, C., Fusaro, A., Gonzales, J.L., Kuiken, T. *et al.* (2023) Avian influenza overview December 2022 – March 2023. European Centre for Disease Prevention and Control, European Union Reference Laboratory for Avian Influenza. *EFSA Journal* 21(3), e07917.

European Food Safety Authority (EFSA), Ståhl, K., Boklund, A., Podgórski, T., Vergne, T. *et al.* (2023) Scientific report on epidemiological analysis of African swine fever in the European Union during 2022. *EFSA Journal* 21(5), 50. DOI: 10.2903/j.efsa.2023.8016.

European Food Safety Authority (EFSA), Ståhl, K., Boklund, A.E., Podgórski, T., Vergne, T. *et al.* (2024) Epidemiological analysis of African swine fever in the European Union during 2023. *EFSA Journal* 22(5), e8809. DOI: 10.2903/j.efsa.2024.8809.

Fafetine, J.M., Coetzee, P., Mubemba, B., Nhambirre, O., Neves, L. *et al.* (2016) Rift Valley fever outbreak in livestock, Mozambique, 2014. *Emerging Infectious Diseases* 22(12), 2165–2167.

Fan, J., Liao, Y., Zhang, M., Liu, C., Li, Z. *et al.* (2021) Anti-classical swine fever virus strategies. *Microorganisms* 9(4), 761.

Feng, J., Hu, Y., Ma, Z., Yu, Q., Zhao, J. *et al.* (2012) Virulent avian infectious bronchitis virus, People's Republic of China. *Emerging Infectious Diseases* 18(12), 1994–2001.

Ferrara, G., D'Anza, E., Rossi, A., Improda, E., Iovane, V. *et al.* (2023) A serological investigation of porcine reproductive and respiratory syndrome and three coronaviruses in the Campania region, Southern Italy. *Viruses* 15(2), 300.

Firestone, S.M., Hayama, Y., Lau, M.S.Y., Yamamoto, T., Nishi, T. *et al.* (2020) Transmission network reconstruction for foot-and-mouth disease outbreaks incorporating farm-level covariates. *PLoS One* 15(7), e0235660.

Fischer, E.A., Boender, G.-J., Nodelijk, G., de Koeijer, A.A. and van Roermund, H.J. (2013) The transmission potential of Rift Valley fever virus among livestock in The Netherlands: A modelling study. *Veterinary Research* 44(1), 58.

Flood, J.S., Porphyre, T., Tildesley, M.J. and Woolhouse, M.E.J. (2013) The performance of approximations of farm contiguity compared to contiguity defined using detailed geographical information in two sample areas in Scotland: Implications for foot-and-mouth disease modelling. *BMC Veterinary Research* 9(1), 198.

Floyd, J.R., Ruktanonchai, N.W., Wardrop, N., Tatem, A.J., Ogola, J. *et al.* (2019) Exploring fine-scale human and livestock movement in western Kenya. *One Health* 7, 100081.

Fountain, J., Woodgate, R., Rast, L. and Hernández-Jover, M. (2018) Assessing biosecurity risks for the introduction and spread of diseases among commercial sheep properties in New South Wales, Australia, using foot-and-mouth disease as a case study. *Frontiers in Veterinary Science* 5, 80.

Fournié, G., Waret-Szkuta, A., Camacho, A., Yigezu, L.M., Pfeiffer, D.U. *et al.* (2018) A dynamic model of transmission and elimination of peste des petits ruminants in Ethiopia. *Proceedings of the National Academy of Sciences of the USA* 115(33), 8454–8459.

Frant, M., Woźniakowski, G. and Pejsak, Z. (2017) African swine fever (ASF) and ticks. No risk of tick-mediated ASF spread in Poland and Baltic states. *Journal of Veterinary Research* 61(4), 375–380.

Franzo, G., Tucciarone, C.M., Moreno, A., Legnardi, M., Massi, P. *et al.* (2020) Phylodynamic analysis and evaluation of the balance between anthropic and environmental factors affecting IBV spreading among Italian poultry farms. *Scientific Reports* 10(1), 7289.

Fritzemeier, J., Teuffert, J., Greiser-Wilke, I., Staubach, C., Schlüter, H. *et al.* (2000) Epidemiology of classical swine fever in Germany in the 1990s. *Veterinary Microbiology* 77(1–2), 29–41.

Gale, P. (2019) Towards a thermodynamic mechanistic model for the effect of temperature on arthropod vector competence for transmission of arboviruses. *Microbial Risk Analysis* 12, 27–43.

Gale, S.B., Miller, G.Y., Eshelman, C.E. and Wells, S.J. (2015) Epidemic simulation of a foot and mouth disease outbreak in Minnesota. *Revue Scientifique et Technique de l'OIE* 34(3), 895–905.

Gallien, S., Moro, A., Lediguerher, G., Catinot, V., Paboeuf, F. *et al.* (2018) Evidence of porcine epidemic diarrhea virus (PEDV) shedding in semen from infected specific pathogen-free boars. *Veterinary Research* 49(1), 7.

Galvis, J.A., Jones, C.M., Prada, J.M., Corzo, C.A. and Machado, G. (2022) The between-farm transmission dynamics of porcine epidemic diarrhoea virus: A short-term forecast modelling comparison and the effectiveness of control strategies. *Transboundary and Emerging Diseases* 69(2), 396–412.

Gao, S., Xu, G., Zeng, Z., Lv, J., Huang, L. *et al.* (2021) Transboundary spread of peste des petits ruminants virus in western China: A prediction model. *PLoS One* 16(9), e0257898.

Gao, Y., Nielsen, L.H., Boklund, A.E., de Jong, M.C.M. and Alban, L. (2022) SWOT analysis of risk factors associated with introduction of African swine fever through vehicles returning after export of pigs. *Frontiers in Veterinary Science* 9, 1049940.

Garner, M.G. and Lack, M.B. (1995) Modelling the potential impact of exotic diseases on regional Australia. *Australian Veterinary Journal* 72(3), 81–87.

Garner, M.G., Hess, G.D. and Yang, X. (2006) An integrated modelling approach to assess the risk of wind-borne spread of foot-and-mouth disease virus from infected premises. *Environmental Modeling & Assessment* 11(3), 195–207.

Garrido-Mantilla, J., Lara, A., Guardado, E., Lopez, J., Nerem, J. *et al.* (2022) Feed or feed transport as a potential route for a porcine epidemic diarrhoea outbreak in a 10,000-sow breeding herd in Mexico. *Transboundary and Emerging Diseases* 69(1), 66–71.

Gaudreault, N.N., Madden, D.W., Wilson, W.C., Trujillo, J.D. and Richt, J.A. (2020) African swine fever virus: An emerging DNA arbovirus. *Frontiers in Veterinary Science* 7, 215.

Gervasi, V. and Guberti, V. (2021) African swine fever endemic persistence in wild boar populations: Key mechanisms explored through modelling. *Transboundary and Emerging Diseases* 68(5), 2812–2825.

Gibson, C.F. and Donaldson, A.I. (1986) Exposure of sheep to natural aerosols of foot-and-mouth disease virus. *Research in Veterinary Science* 41(1), 45–49.

Gilbert, M. and Pfeiffer, D.U. (2012) Risk factor modelling of the spatio-temporal patterns of highly pathogenic avian influenza (HPAIV) H5N1: A review. *Spatial and Spatio-Temporal Epidemiology* 3(3), 173–183.

Gloster, J., Champion, H.J., Sørensen, J.H., Mikkelsen, T., Ryall, D.B. *et al.* (2003) Airborne transmission of foot-and-mouth disease virus from Burnside Farm, Heddon-on-the-Wall, Northumberland, during the 2001 epidemic in the United Kingdom. *Veterinary Record* 152(17), 525–533.

Gloster, J., Jones, A., Redington, A., Burgin, L., Sørensen, J.H. *et al.* (2010) Airborne spread of foot-and-mouth disease – model intercomparison. *Veterinary Journal* 183(3), 278–286.

Golnar, A.J., Kading, R.C. and Hamer, G.L. (2018) Quantifying the potential pathways and locations of Rift Valley fever virus entry into the United States. *Transboundary and Emerging Diseases* 65(1), 85–95.

Golnar, A.J., Martin, E., Wormington, J.D., Kading, R.C., Teel, P.D. *et al.* (2019) Reviewing the potential vectors and hosts of African swine fever virus transmission in the United States. *Vector Borne and Zoonotic Diseases* 19(7), 512–524.

Grace, K.E.F., Papadopoulou, C., Floyd, T., Avigad, R., Collins, S. *et al.* (2020) Risk-based surveillance for bluetongue virus in cattle on the south coast of England in 2017 and 2018. *Veterinary Record* 187(11), e96.

Green, D.M., Kiss, I.Z. and Kao, R.R. (2006) Modelling the initial spread of foot-and-mouth disease through animal movements. *Proceedings. Biological Sciences* 273(1602), 2729–2735.

Greer, A.L., Spence, K. and Gardner, E. (2017) Understanding the early dynamics of the 2014 porcine epidemic diarrhea virus (PEDV) outbreak in Ontario using the incidence decay and exponential adjustment (IDEA) model. *BMC Veterinary Research* 13(1), 8.

Gubbins, S., Carpenter, S., Baylis, M., Wood, J.L.N. and Mellor, P.S. (2008) Assessing the risk of bluetongue to UK livestock: Uncertainty and sensitivity analyses of a temperature-dependent model for the basic reproduction number. *Journal of the Royal Society, Interface* 5(20), 363–371.

Gubbins, S. (2019) Using the basic reproduction number to assess the risk of transmission of lumpy skin disease virus by biting insects. *Transboundary and Emerging Diseases* 66(5), 1873–1883. DOI: 10.1111/tbed.13216.

Guinat, C., Gogin, A., Blome, S., Keil, G., Pollin, R. *et al.* (2016) Transmission routes of African swine fever virus to domestic pigs: Current knowledge and future research directions. *Veterinary Record* 178(11), 262–267.

Guinat, C., Porphyre, T., Gogin, A., Dixon, L., Pfeiffer, D.U. *et al.* (2018) Inferring within-herd transmission parameters for African swine fever virus using mortality data from outbreaks in the Russian Federation. *Transboundary and Emerging Diseases* 65(2), e264–e271.

Guinat, C., Valenzuela Agüí, C., Vaughan, T.G., Scire, J., Pohlmann, A. *et al.* (2022) Disentangling the role of poultry farms and wild birds in the spread of highly pathogenic avian influenza virus in Europe. *Virus Evolution* 8(2), veac073.

Hadorn, D.C., Racloz, V., Schwermer, H. and Stärk, K.D.C. (2009) Establishing a cost-effective national surveillance system for bluetongue using scenario tree modelling. *Veterinary Research* 40(6), 57.

Haesendonck, R., Verlinden, M., Devos, G., Michiels, T., Butaye, P. *et al.* (2014) High seroprevalence of respiratory pathogens in hobby poultry. *Avian Diseases* 58(4), 623–627.

Hagerman, A.D., South, D.D., Sondgerath, T.C., Patyk, K.A., Sanson, R.L. *et al.* (2018) Temporal and geographic distribution of weather conditions favorable to airborne spread of foot-and-mouth disease in the coterminous United States. *Preventive Veterinary Medicine* 161, 41–49.

Hakizimana, J.N., Yona, C., Kamana, O., Nauwynck, H. and Misinzo, G. (2021) African swine fever virus circulation between Tanzania and neighboring countries: A systematic review and meta-analysis. *Viruses* 13(2), 306.

Hakobyan, S.A., Ross, P.A., Bayramyan, N.V., Poghosyan, A.A., Avetisyan, A.S. *et al.* (2022) Experimental models of ecological niches for African swine fever virus. *Veterinary Microbiology* 266, 109365.

Halasa, T., Boklund, A., Bøtner, A., Toft, N. and Thulke, H.H. (2016a) Simulation of spread of African Swine fever, including the effects of residues from dead animals. *Frontiers in Veterinary Science* 3, 6.

Halasa, T., Bøtner, A., Mortensen, S., Christensen, H., Toft, N. *et al.* (2016b) Simulating the epidemiological and economic effects of an African swine fever epidemic in industrialized swine populations. *Veterinary Microbiology* 193, 7–16.

Halecker, S., Joseph, S., Mohammed, R., Wernery, U., Mettenleiter, T.C. *et al.* (2020) Comparative evaluation of different antigen detection methods for the detection of peste des petits ruminants virus. *Transboundary and Emerging Diseases* 67(6), 2881–2891.

Hamdi, J., Bamouh, Z., Jazouli, M., Alhyane, M., Safini, N. *et al.* (2021) Experimental infection of indigenous north African goats with Goatpox virus. *Acta Veterinaria Scandinavica* 63(1), 9.

Hammami, P. (2016) Peste des petits ruminants en Afrique subsaharienne: Modélisation et analyse des stratégies de vaccination dans un contexte de bien public mondial. Doctoral thesis, Université de Montpellier.

Hanke, D., Pohlmann, A., Sauter-Louis, C., Höper, D., Stadler, J. *et al.* (2017) Porcine epidemic diarrhea in Europe: In-detail analyses of disease dynamics and molecular epidemiology. *Viruses* 9(7), 177.

Hartley, D.M., Rinderknecht, J.L., Nipp, T.L., Clarke, N.P., Snowder, G.D. *et al.* (2011) Potential effects of Rift Valley fever in the United States. *Emerging Infectious Diseases* 17(8), e1.

Hautefeuille, C., Dauphin, G. and Peyre, M. (2020) Knowledge and remaining gaps on the role of animal and human movements in the poultry production and trade networks in the global spread of avian influenza viruses – a scoping review. *PLoS One* 15(3), e0230567.

Hernández-Jover, M., Schembri, N., Holyoake, P.K., Toribio, J.-A.L.M.L. and Martin, P.A.J. (2016) A comparative assessment of the risks of introduction and spread of foot-and-mouth disease among different pig sectors in Australia. *Frontiers in Veterinary Science* 3, 85.

Herrera-Diestra, J.L., Tildesley, M., Shea, K. and Ferrari, M.J. (2022) Cattle transport network predicts endemic and epidemic foot-and-mouth disease risk on farms in Turkey. *PLoS Computational Biology* 18(8), e1010354.

Herzog, C.M., de Glanville, W.A., Willett, B.J., Cattadori, I.M., Kapur, V. *et al.* (2020) *Peste des petits ruminants* virus transmission scaling and husbandry practices that contribute to increased transmission risk: An investigation among sheep, goats, and cattle in northern Tanzania. *Viruses* 12(9), 930.

Heuer, C., French, N.P., Jackson, R. and Mackereth, G.F. (2007) Application of modelling to determine the absence of foot-and-mouth disease in the face of a suspected incursion. *New Zealand Veterinary Journal* 55(6), 289–296.

Hoffmann, B., Wiesner, H., Maltzan, J., Mustefa, R., Eschbaumer, M. *et al.* (2012) Fatalities in wild goats in Kurdistan associated with peste des petits ruminants virus. *Transboundary and Emerging Diseases* 59(2), 173–176.

Hoinville, L.J., Alban, L., Drewe, J.A., Gibbens, J.C., Gustafson, L. *et al.* (2013) Proposed terms and concepts for describing and evaluating animal-health surveillance systems. *Preventive Veterinary Medicine* 112(1–2), 1–12.

Hone, J., Pech, R. and Yip, P. (1992) Estimation of the dynamics and rate of transmission of classical swine fever (hog cholera) in wild pigs. *Epidemiology and Infection* 108(2), 377–386.

Howey, R., Bankowski, B., Juleff, N., Savill, N.J., Gibson, D. *et al.* (2012) Modelling the within-host dynamics of the foot-and-mouth disease virus in cattle. *Epidemics* 4(2), 93–103.

Hunter, P. (2022) Europe's worst ever bird flu outbreak: This year's epidemic of highly pathogenic avian flu has had a devastating impact on wild and domestic birds and severe economic consequences. *EMBO Reports* 23(10), e56048.

Ivorra, B., Martínez-López, B., Sánchez-Vizcaíno, J.M. and Ramos, Á.M. (2014) Mathematical formulation and validation of the be-FAST model for classical swine fever virus spread between and within farms. *Annals of Operations Research* 219(1), 25–47.

Jamil, T., Akar, K., Erdenlig, S., Murugaiyan, J., Sandalakis, V. *et al.* (2022) Spatio-temporal distribution of brucellosis in European terrestrial and marine wildlife species and its regional implications. *Microorganisms* 10(10), 1970.

Jauniaux, T.P., De Clercq, K.E., Cassart, D.E., Kennedy, S., Vandenbussche, F.E. *et al.* (2008) Bluetongue in Eurasian lynx. *Emerging Infectious Diseases* 14(9), 1496–1498.

Jung, K. and Saif, L.J. (2015) Porcine epidemic diarrhea virus infection: Etiology, epidemiology, pathogenesis and immunoprophylaxis. *Veterinary Journal* 204(2), 134–143.

Jurado, C., Martínez-Avilés, M., de la Torre, A., Štukelj, M., de Carvalho Ferreira, H.C. *et al.* (2018) Relevant measures to prevent the spread of African swine fever in the European Union domestic pig sector. *Frontiers in Veterinary Science* 5, 77.

Kaden, V., Steyer, H., Schnabel, J. and Bruer, W. (2005) Classical swine fever (CSF) in wild boar: The role of the transplacental infection in the perpetuation of CSF. *Journal of Veterinary Medicine, Series B* 52(4), 161–164.

Kao, R.R., Danon, L., Green, D.M. and Kiss, I.Z. (2006) Demographic structure and pathogen dynamics on the network of livestock movements in Great Britain. *Proceedings. Biological Sciences* 273(1597), 1999–2007.

Kardjadj, M. (2017) Prevalence, distribution, and risk factor for sheep pox and goat pox (SPGP) in Algeria. *Tropical Animal Health and Production* 49(3), 649–652.

Karte, C., Platje, N., Bullermann, J., Beer, M., Höper, D. *et al.* (2020) Re-emergence of porcine epidemic diarrhea virus in a piglet-producing farm in northwestern Germany in 2019. *BMC Veterinary Research* 16(1), 329.

Kasari, T.R., Carr, D.A., Lynn, T.V. and Weaver, J.T. (2008) Evaluation of pathways for release of Rift Valley fever virus into domestic ruminant livestock, ruminant wildlife, and human populations in the continental United States. *Journal of the American Veterinary Medical Association* 232(4), 514–529.

Katsiolis, A., Papanikolaou, E., Stournara, A., Giakkoupi, P., Papadogiannakis, E. *et al.* (2022) Molecular detection of *Brucella* spp. in ruminant herds in Greece. *Tropical Animal Health and Production* 54(3), 173.

Keeling, M.J. (2005) Models of foot-and-mouth disease. *Proceedings. Biological Sciences* 272(1569), 1195–1202.

Kenawy, M.A., Abdel-Hamid, Y.M. and Beier, J.C. (2018) Rift Valley fever in Egypt and other African countries: Historical review, recent outbreaks and possibility of disease occurrence in Egypt. *Acta Tropica* 181, 40–49.

Khanal, P., Olcha, M. and Niederwerder, M.C. (2022) Detection of African swine fever virus in feed dust collected from experimentally inoculated complete feed using quantitative PCR and virus titration assays. *Transboundary and Emerging Diseases* 69(1), 97–102.

Khera, H., Samad, A., Abbas, A., Mehtab, U. and Rehman, A. (2022) Diagnosis, prevention and control strategies of infectious bronchitis virus. *Science Letters* 10(1), 16–20.

Khurana, S.K., Sehrawat, A., Tiwari, R., Prasad, M., Gulati, B. *et al.* (2021) Bovine brucellosis – a comprehensive review. *Veterinary Quarterly* 41(1), 61–88.

Kim, Y., Yang, M., Goyal, S.M., Cheeran, M.C. and Torremorell, M. (2017) Evaluation of biosecurity measures to prevent indirect transmission of porcine epidemic diarrhea virus. *BMC Veterinary Research* 13(1), 89.

Kim, Y., Métras, R., Dommergues, L., Youssouffi, C., Combo, S. *et al.* (2021) The role of livestock movements in the spread of Rift Valley fever virus in animals and humans in Mayotte, 2018–19. *PLoS Neglected Tropical Diseases* 15(3), e0009202.

Kirkeby, C. and Ward, M.P. (2022) A review of estimated transmission parameters for the spread of avian influenza viruses. *Transboundary and Emerging Diseases* 69(6), 3238–3246.

Kitching, R.P. and Mellor, P.S. (1986) Insect transmission of capripoxvirus. *Research in Veterinary Science* 40(2), 255–258.

Kitching, R.P. and Hughes, G.J. (2002) Clinical variation in foot and mouth disease: Sheep and goats. *Revue Scientifique et Technique de l'OIE* 21(3), 505–512.

Kitching, R.P. and Taylor, W.P. (1985) Clinical and antigenic relationship between isolates of sheep and goat pox viruses. *Tropical Animal Health and Production* 17(2), 64–74. DOI: 10.1007/BF02360774.

Kitching, R.P., Hutber, A.M. and Thrusfield, M.V. (2005) A review of foot-and-mouth disease with special consideration for the clinical and epidemiological factors relevant to predictive modelling of the disease. *Veterinary Journal* 169(2), 197–209.

Kivaria, F.M., Kwiatek, O., Kapaga, A.M., Swai, E.S., Libeau, G. *et al.* (2013) The incursion, persistence and spread of peste des petits ruminants in Tanzania: Epidemiological patterns and predictions. *Onderstepoort Journal of Veterinary Research* 80(1), 1–10.

Klausner, Z., Klement, E. and Fattal, E. (2015) Modeling long distance dispersal of airborne foot-and-mouth disease virus as a polydisperse aerosol – application to the emergence of a new strain from Egypt to Israel. *Atmospheric Environment* 122, 332–342.

Kleiboeker, S.B., Burrage, T.G., Scoles, G.A., Fish, D. and Rock, D.L. (1998) African swine fever virus infection in the argasid host, *Ornithodoros porcinus porcinus*. *Journal of Virology* 72(3), 1711–1724.

Klement, E., Broglia, A., Antoniou, S.E., Tsiamadis, V., Plevraki, E. *et al.* (2020) Neethling vaccine proved highly effective in controlling lumpy skin disease epidemics in the Balkans. *Preventive Veterinary Medicine* 181, 104595.

Klinkenberg, D., de Bree, J., Laevens, H. and de Jong, M.C.M. (2002) Within- and between-pen transmission of classical swine fever virus: A new method to estimate the basic reproduction ratio from transmission experiments. *Epidemiology and Infection* 128(2), 293–299.

Knight-Jones, T.J., Gubbins, S., Bulut, A.N., Stärk, K.D., Pfeiffer, D.U. *et al.* (2016) Mass vaccination, immunity and coverage: Modelling population protection against foot-and-mouth disease in Turkish cattle. *Scientific Reports* 6, 22121.

Knight-Jones, T.J.D. and Rushton, J. (2013) The economic impacts of foot and mouth disease – what are they, how big are they and where do they occur? *Preventive Veterinary Medicine* 112(3–4), 161–173.

Korennoy, F., Gulenkin, V., Gogin, A., Vergne, T. and Karaulov, A. (2017) Estimating the basic reproductive number for African swine fever using the Ukrainian historical epidemic of 1977. *Transboundary and Emerging Diseases* 64(6), 1858–1866.

Kroeker, A.L., Babiuk, S., Pickering, B.S., Richt, J.A. and Wilson, W.C. (2020) Livestock challenge models of Rift Valley fever for agricultural vaccine testing. *Frontiers in Veterinary Science* 7, 238.

Kumar, N., Maherchandani, S., Kashyap, S.K., Singh, S.V., Sharma, S. *et al.* (2014) Peste des petits ruminants virus infection of small ruminants: A comprehensive review. *Viruses* 6(6), 2287–2327.

Laevens, H., Koenen, F., Deluyker, H. and de Kruif, A. (1999) Experimental infection of slaughter pigs with classical swine fever virus: Transmission of the virus, course of the disease and antibody response. *Veterinary Record* 145(9), 243–248.

Lafaye, M., Sall, B., Ndiaye, Y., Vignolles, C., Tourre, Y.M. *et al.* (2013) Rift Valley fever dynamics in Senegal: A project for pro-active adaptation and improvement of livestock raising management. *Geospatial Health* 8(1), 279–288.

Lagare, A., Fall, G., Ibrahim, A., Ousmane, S., Sadio, B. *et al.* (2019) First occurrence of Rift Valley fever outbreak in Niger, 2016. *Veterinary Medicine and Science* 5(1), 70–78.

Lambin, E.F., Tran, A., Vanwambeke, S.O., Linard, C. and Soti, V. (2010) Pathogenic landscapes: Interactions between land, people, disease vectors, and their animal hosts. *International Journal of Health Geographics* 9, 54.

Lee, C. (2015) Porcine epidemic diarrhea virus: An emerging and re-emerging epizootic swine virus. *Virology Journal* 12, 193.

Lee, D.H., Criado, M.F. and Swayne, D.E. (2021) Pathobiological origins and evolutionary history of highly pathogenic avian influenza viruses. *Cold Spring Harbor Perspectives in Medicine* 11(2), a038679.

Leedale, J., Jones, A.E., Caminade, C. and Morse, A.P. (2016) A dynamic, climate-driven model of Rift Valley fever. *Geospatial Health* 11(1 Suppl), 394.

Liang, G. and Fang, H. (2022) Transmission dynamics of a brucellosis model with incubation period. *International Journal of Biomathematics* 15(5).

Liang, J., Zhao, Z. and Li, C. (2020) Rich dynamics of a brucellosis model with transport. *Complexity* 2020, 1–7.

Linden, A., Licoppe, A., Volpe, R., Paternostre, J., Lesenfants, C. *et al.* (2019) Summer 2018: African swine fever virus hits north-western Europe. *Transboundary and Emerging Diseases* 66(1), 54–55.
Liu, H., Ren, Y., Wang, T., Shan, H. and Wong, K.W. (2023) Fuzzy model for quantitative assessment of the epidemic risk of African swine fever within Australia. *Preventive Veterinary Medicine* 213, 105884.
Liu, Q. and Wang, H.Y. (2021) Porcine enteric coronaviruses: An updated overview of the pathogenesis, prevalence, and diagnosis. *Veterinary Research Communications* 45, 75–86. DOI: 10.1007/s11259-021-09808-0.
Liu, W., Wu, X., Wang, Z., Bao, J., Li, L. *et al.* (2013) Virus excretion and antibody dynamics in goats inoculated with a field isolate of peste des petits ruminants virus. *Transboundary and Emerging Diseases* 60(Suppl 2), 63–68.
Lombardi, M.E., Ladman, B.S., Alphin, R.L. and Benson, E.R. (2008) Inactivation of avian influenza virus using common detergents and chemicals. *Avian Diseases* 52(1), 118–123.
Longjam, N., Deb, R., Sarmah, A.K., Tayo, T., Awachat, V.B. *et al.* (2011) A brief review on diagnosis of foot-and-mouth disease of livestock: Conventional to molecular tools. *Veterinary Medicine International* 2011, 905768.
Longworth, N., Mourits, M.C.M. and Saatkamp, H.W. (2014) Economic analysis of HPAI control in The Netherlands I: Epidemiological modelling to support economic analysis. *Transboundary and Emerging Diseases* 61(3), 199–216.
Lorenzi, G., Borella, L., Alborali, G.L., Prodanov-Radulović, J., Štukelj, M. *et al.* (2020) African swine fever: A review of cleaning and disinfection procedures in commercial pig holdings. *Research in Veterinary Science* 132, 262–267.
Lou, P., Wang, L., Zhang, X., Xu, J. and Wang, K. (2016) Modelling seasonal brucellosis epidemics in Bayingolin Mongol autonomous prefecture of Xinjiang, China, 2010–2014. *BioMed Research International* 2016, 5103718.
Lowe, J., Gauger, P., Harmon, K., Zhang, J., Connor, J. *et al.* (2014) Role of transportation in spread of porcine epidemic diarrhea virus infection, United States. *Emerging Infectious Diseases* 20(5), 872–874.
Lu, Z., Yu, S., Wang, W., Chen, W., Wang, X. *et al.* (2022) Development of foot-and-mouth disease vaccines in recent years. *Vaccines* 10(11), 1817.
Machavarapu, M., Poonati, R., Mallepaddi, P.C., Gundlamadugu, V.V., Raghavendra, S. *et al.* (2019) Endemic brucellosis in Indian animal and human populations: A billion dollar issue. *Current Trends in Biotechnology and Pharmacy* 13(2), 112–123.
Magori-Cohen, R., Louzoun, Y., Herziger, Y., Oron, E., Arazi, A. *et al.* (2012) Mathematical modelling and evaluation of the different routes of transmission of lumpy skin disease virus. *Veterinary Research* 43, 1–13.
Mai, T.N., Nguyen, T.T., Vu, V.A., Vu, T.N. and Huynh, T.M.L. (2022a) Estimation of the herd-level basic reproduction number for African swine fever in Vietnam, 2019. *Veterinary World* 15(12), 2850–2855.
Mai, T.N., Sekiguchi, S., Huynh, T.M., Cao, T.B., Le, V.P. *et al.* (2022b) Dynamic models of within-herd transmission and recommendation for vaccination coverage requirement in the case of African swine fever in Vietnam. *Veterinary Sciences* 9(6), 292.
Makita, K. and Yamamoto, T. (2019) Basic reproduction number as a measure of the rapidity of the inter-farm spread of porcine epidemic diarrhea during the initial phase of the epidemic in Japan in 2013–2014. *Journal of Veterinary Epidemiology* 23(2), 111–118.
Maksyutov, R.A., Gavrilova, E.V., Agafonov, A.P., Taranov, O.S., Glotov, A.G. *et al.* (2015) An outbreak of sheep pox in Zabajkalskij Kray of Russia. *Transboundary and Emerging Diseases* 62(4), 453–456.
Malek, A. and Hoque, A. (2022) Mathematical modeling of bird flu with vaccination and treatment for the poultry farms. *Comparative Immunology, Microbiology and Infectious Diseases* 80, 101721.
Malesios, C., Demiris, N., Abas, Z., Dadousis, K. and Koutroumanidis, T. (2014) Modeling sheep pox disease from the 1994–1998 epidemic in Evros Prefecture, Greece. *Spatial and Spatio-Temporal Epidemiology* 11, 1–10.
Malesios, C., Chatzipanagiotou, M., Demiris, N., Kantartzis, A. and Chatzilazarou, G. (2020) A quantitative analysis of the spatial and temporal evolution patterns of the bluetongue virus outbreak in the island of Lesvos, Greece, in 2014. *Transboundary and Emerging Diseases*.
Malladi, S., Ssematimba, A., Bonney, P.J., St Charles, K.M., Boyer, T. *et al.* (2022) Predicting the time to detect moderately virulent African swine fever virus in finisher swine herds using a stochastic disease transmission model. *BMC Veterinary Research* 18(1), 84.

Mangana-Vougiouka, O., Kottaridi, C. and Nomikou, K. (2008) The epidemiology of sheep pox in Greece from 1987 to 2007. *Revue Scientifique et Technique de l'OIE* 27(3), 899–905.

Manić, M., Stojiljković, M., Petrović, M., Nišavić, J., Bacić, D. *et al.* (2019) Epizootic features and control measures for lumpy skin disease in south-east Serbia in 2016. *Transboundary and Emerging Diseases* 66(5), 2087–2099.

Martínez-López, B., Perez, A.M., De la Torre, A. and Sánchez-Vizcaíno Rodriguez, J.M. (2008) Quantitative risk assessment of foot-and-mouth disease introduction into Spain via importation of live animals. *Preventive Veterinary Medicine* 86(1–2), 43–56.

Martínez-López, B., Ivorra, B., Ngom, D., Ramos, A.M. and Sánchez-Vizcaíno, J.M. (2012) A novel spatial and stochastic model to evaluate the within and between farm transmission of classical swine fever virus: II validation of the model. *Veterinary Microbiology* 155(1), 21–32.

Mayo, C.E., Crossley, B.M., Hietala, S.K., Gardner, I.A., Breitmeyer, R.E. *et al.* (2010) Colostral transmission of bluetongue virus nucleic acid among newborn dairy calves in California. *Transboundary and Emerging Diseases* 57(4), 277–281.

Mellor, P.S. and Boorman, J. (1995) The transmission and geographical spread of African horse sickness and bluetongue viruses. *Annals of Tropical Medicine & Parasitology* 89(1), 1–15.

Meltzer, E., Sidi, Y., Smolen, G., Banai, M., Bardenstein, S. *et al.* (2010) Sexually transmitted brucellosis in humans. *Clinical Infectious Diseases* 51(2), e12–15.

Menzies, F.D., McCullough, S.J., McKeown, I.M., Forster, J.L., Jess, S. *et al.* (2008) Evidence for transplacental and contact transmission of bluetongue virus in cattle. *Veterinary Record* 163(7), 203–209.

Métras, R., Cavalerie, L., Dommergues, L., Mérot, P., Edmunds, W.J. *et al.* (2016) The epidemiology of Rift Valley fever in Mayotte: Insights and perspectives from 11 years of data. *PLoS Neglected Tropical Diseases* 10(6), e0004783.

Métras, R., Fournié, G., Dommergues, L., Camacho, A., Cavalerie, L. *et al.* (2017) Drivers for Rift Valley fever emergence in Mayotte: A Bayesian modelling approach. *PLoS Neglected Tropical Diseases* 11(7), e0005767.

Miguel, E., Grosbois, V., Caron, A., Boulinier, T., Fritz, H. *et al.* (2013) Contacts and foot and mouth disease transmission from wild to domestic bovines in Africa. *Ecosphere* 4(4), 1–32.

Miller, B.J. (2022) Why unprecedented bird flu outbreaks sweeping the world are concerning scientists. *Nature* 606, 18–19.

Miller, M. (2009) Analyse qualitative du risque d'introduction de la peste des petits ruminants en france métropolitaine à partir du pourtour méditerranéen. Available at: https://eurl-ppr.cirad.fr/FichiersComplementaires/PPR-ENG/res/miller_2009.pdf (accessed 11 December 2024).

Miller, G.Y., Gale, S.B., Eshelman, C.E. and Wells, S.J. (2015) Emergency vaccination use in a modelled foot and mouth disease outbreak in Minnesota. *Revue Scientifique et Technique de l'OIE* 34(3), 729–740.

Moennig, V., Floegel-Niesmann, G. and Greiser-Wilke, I. (2003) Clinical signs and epidemiology of classical swine fever: A review of new knowledge. *Veterinary Journal* 165(1), 11–20.

Moennig, V. (2015) The control of classical swine fever in wild boar. *Frontiers in Microbiology* 6, 1211. DOI: 10.3389/fmicb.2015.01211.

Möhlmann, T.W.R., Keeling, M.J., Wennergren, U., Favia, G., Santman-Berends, I. *et al.* (2021) Biting midge dynamics and bluetongue transmission: A multiscale model linking catch data with climate and disease outbreaks. *Scientific Reports* 11(1), 1892.

Möller, J., Moritz, T., Schlottau, K., Krstevski, K., Hoffmann, D. *et al.* (2019) Experimental lumpy skin disease virus infection of cattle: Comparison of a field strain and a vaccine strain. *Archives of Virology* 164, 2931–2941.

Mosomtai, G., Evander, M., Sandström, P., Ahlm, C., Sang, R. *et al.* (2016) Association of ecological factors with Rift Valley fever occurrence and mapping of risk zones in Kenya. *International Journal of Infectious Diseases* 46, 49–55.

Mosomtai, G., Evander, M., Mundia, C., Sandström, P., Ahlm, C. *et al.* (2018) Datasets for mapping pastoralist movement patterns and risk zones of Rift Valley fever occurrence. *Data in Brief* 16, 762–770.

Motta, P., Garner, G., Hòvari, M., Alexandrov, T., Bulut, A. *et al.* (2019) A framework for reviewing livestock disease reporting systems in high-risk areas: assessing performance and perceptions towards foot and mouth disease reporting in the Thrace region of Greece. *Transboundary and Emerging Diseases* 66(3), 1268–1279.

Murai, K., Moriguchi, S., Hayama, Y., Kobayashi, S., Miyazaki, A. *et al.* (2018) Mathematical modeling of porcine epidemic diarrhea virus dynamics within a farrow-to-finish swine farm to investigate the effects of control measures. *Preventive Veterinary Medicine* 149, 115–124.
Mutua, F. and Dione, M. (2021) The context of application of biosecurity for control of African swine fever in smallholder pig systems. *Frontiers in Veterinary Science* 8, 689811.
Naeem, M., Alahmed, A.M., Kheir, S.M. and Sallam, M.F. (2016) Spatial distribution modeling of *Stegomyia aegypti* and *Culex tritaeniorhynchus* (Diptera: Culicidae) in Al-Bahah Province, Kingdom of Saudi Arabia. *Tropical Biomedicine* 33(2), 295–310.
Namazi, F. and Khodakaram Tafti, A. (2021) Lumpy skin disease, an emerging transboundary viral disease: A review. *Veterinary Medicine and Science* 7(3), 888–896.
Nelson, N., Paton, D.J., Gubbins, S., Colenutt, C., Brown, E. *et al.* (2017) Predicting the ability of preclinical diagnosis to improve control of farm-to-farm foot-and-mouth disease transmission in cattle. *Journal of Clinical Microbiology* 55(6), 1671–1681.
Neumann, E.J., Hall, W.F., Dahl, J., Hamilton, D. and Kurian, A. (2021) Is transportation a risk factor for African swine fever transmission in Australia? A review. *Australian Veterinary Journal* 99(11), 459–468.
Nie, J., Sun, G.-Q., Sun, X.-D., Zhang, J., Wang, N.A.N. *et al.* (2014) Modeling the transmission dynamics of dairy cattle brucellosis in Jilin Province, China. *Journal of Biological Systems* 22(4), 533–554.
Nielsen, J.P., Larsen, T.S., Halasa, T. and Christiansen, L.E. (2017) Estimation of the transmission dynamics of African swine fever virus within a swine house. *Epidemiology and Infection* 145(13), 2787–2796.
Njeumi, F., Bailey, D., Soula, J.J., Diop, B. and Tekola, B.G. (2020) Eradicating the scourge of peste des petits ruminants from the world. *Viruses* 12(3), 313.
Nöremark, M., Håkansson, N., Lindström, T., Wennergren, U. and Lewerin, S.S. (2009) Spatial and temporal investigations of reported movements, births and deaths of cattle and pigs in Sweden. *Acta Veterinaria Scandinavica* 51(1), 37.
Nuanualsuwan, S., Songkasupa, T., Boonpornprasert, P., Suwankitwat, N., Lohlamoh, W. *et al.* (2022) Persistence of African swine fever virus on porous and non-porous fomites at environmental temperatures. *Porcine Health Management* 8(1), 34.
Nyerere, N., Luboobi, L.S., Mpeshe, S.C. and Shirima, G.M. (2020) Modeling the impact of seasonal weather variations on the infectiology of brucellosis. *Computational and Mathematical Methods in Medicine* 2020, 8972063.
Oberin, M., Hillman, A., Ward, M.P., Holley, C., Firestone, S. *et al.* (2022) The potential role of wild suids in African swine fever spread in Asia and the Pacific region. *Viruses* 15(1), 61.
Ochieng, A.O., Nanyingi, M., Kipruto, E., Ondiba, I.M., Amimo, F.A. *et al.* (2016) Ecological niche modelling of Rift Valley fever virus vectors in Baringo, Kenya. *Infection Ecology & Epidemiology* 6, 32322.
OIE (2020). Available at: https://www.vlaanderen.be/vlam/en/press/oie-confirms-belgium-has-regained-its-asf-free-status (accessed 15 December 2024).
Olesen, A.S., Belsham, G.J., Bruun Rasmussen, T., Lohse, L., Bødker, R. *et al.* (2020) Potential routes for indirect transmission of African swine fever virus into domestic pig herds. *Transboundary and Emerging Diseases* 67(4), 1472–1484.
Olive, M.M., Chevalier, V., Grosbois, V., Tran, A., Andriamandimby, S.-F. *et al.* (2016) Integrated analysis of environment, cattle and human serological data: Risks and mechanisms of transmission of Rift Valley fever in Madagascar. *PLoS Neglected Tropical Diseases* 10(7), e0004827.
Olive, M.M., Grosbois, V., Tran, A., Nomenjanahary, L.A., Rakotoarinoro, M. *et al.* (2017) Reconstruction of Rift Valley fever transmission dynamics in Madagascar: Estimation of force of infection from seroprevalence surveys using Bayesian modelling. *Scientific Reports* 7(1), 39870.
Orsel, K., Bouma, A., Dekker, A., Stegeman, J.A. and De Jong, M.C.M. (2009) Foot and mouth disease virus transmission during the incubation period of the disease in piglets, lambs, calves, and dairy cows. *Preventive Veterinary Medicine* 88(2), 158–163.
Ozkul, A., Akca, Y., Alkan, F., Barrett, T., Karaoglu, T. *et al.* (2002) Prevalence, distribution, and host range of peste des petits ruminants virus, Turkey. *Emerging Infectious Diseases* 8(7), 708–712.
Pacheco, J.M., Arzt, J. and Rodriguez, L.L. (2010) Early events in the pathogenesis of foot-and-mouth disease in cattle after controlled aerosol exposure. *Veterinary Journal* 183(1), 46–53.
Parida, S., Muniraju, M., Mahapatra, M., Muthuchelvan, D., Buczkowski, H. *et al.* (2015) Peste des petits ruminants. *Veterinary Microbiology* 181(1–2), 90–106.
Parida, S., Selvaraj, M., Gubbins, S., Pope, R., Banyard, A. *et al.* (2019) Quantifying levels of peste des petits ruminants (PPR) virus in excretions from experimentally infected goats and its importance for nascent PPR eradication programme. *Viruses* 11(3), 249.

Pasick, J., Berhane, Y., Ojkic, D., Maxie, G., Embury-Hyatt, C. *et al.* (2014) Investigation into the role of potentially contaminated feed as a source of the first-detected outbreaks of porcine epidemic diarrhea in Canada. *Transboundary and Emerging Diseases* 61(5), 397–410.

Paton, D.J., Gubbins, S. and King, D.P. (2018) Understanding the transmission of foot-and-mouth disease virus at different scales. *Current Opinion in Virology* 28, 85–91.

Penrith, M.-L., van Heerden, J., Pfeiffer, D.U., Oļševskis, E., Depner, K. *et al.* (2023) Innovative research offers new hope for managing African swine fever better in resource-limited smallholder farming settings: A timely update. *Pathogens* 12(2), 355.

Perri, A.M., Poljak, Z., Dewey, C., Harding, J.C.S. and O'Sullivan, T.L. (2018) An epidemiological investigation of the early phase of the porcine epidemic diarrhea (PED) outbreak in Canadian swine herds in 2014: A case–control study. *Preventive Veterinary Medicine* 150, 101–109.

Petrov, A., Blohm, U., Beer, M., Pietschmann, J. and Blome, S. (2014) Comparative analyses of host responses upon infection with moderately virulent classical swine fever virus in domestic pigs and wild boar. *Virology Journal* 11, 134.

Pietschmann, J., Guinat, C., Beer, M., Pronin, V., Tauscher, K. *et al.* (2015) Course and transmission characteristics of oral low-dose infection of domestic pigs and European wild boar with a Caucasian African swine fever virus isolate. *Archives of Virology* 160(7), 1657–1667.

Postel, A., Moennig, V. and Becher, P. (2013) Classical swine fever in Europe – the current situation. *Berliner Und Munchener Tierarztliche Wochenschrift* 126(11–12), 468–475.

Postel, A., Meyer, D., Petrov, A. and Becher, P. (2017) Recent emergence of a novel porcine pestivirus: Interference with classical swine fever diagnosis? *Emerging Microbes & Infections* 6(4), e19.

Purse, B.V., Tatem, A.J., Caracappa, S., Rogers, D.J., Mellor, P.S. *et al.* (2004) Modelling the distributions of *Culicoides* bluetongue virus vectors in Sicily in relation to satellite-derived climate variables. *Medical and Veterinary Entomology* 18(2), 90–101.

Ramakrishnan, S. and Kappala, D. (2019) Avian infectious bronchitis virus. *Recent Advances in Animal Virology* 301–319.

Rasmussen, L.D., Savini, G., Lorusso, A., Bellacicco, A., Palmarini, M. *et al.* (2013) Transplacental transmission of field and rescued strains of BTV-2 and BTV-8 in experimentally infected sheep. *Veterinary Research* 44(1), 1–15.

Ravaomanana, J., Michaud, V., Jori, F., Andriatsimahavandy, A., Roger, F. *et al.* (2010) First detection of African swine fever virus in *Ornithodoros porcinus* in Madagascar and new insights into tick distribution and taxonomy. *Parasites & Vectors* 3(1), 1–9.

Reddy, V.R.A.P., Trus, I., Desmarets, L.M.B., Li, Y., Theuns, S. *et al.* (2016) Productive replication of nephropathogenic infectious bronchitis virus in peripheral blood monocytic cells, a strategy for viral dissemination and kidney infection in chickens. *Veterinary Research* 47(1), 70.

Renken, C., Nathues, C., Swam, H., Fiebig, K., Weiss, C. *et al.* (2021) Application of an economic calculator to determine the cost of porcine reproductive and respiratory syndrome at farm-level in 21 pig herds in Germany. *Porcine Health Management* 7(1), 3.

Ribbens, S., Dewulf, J., Koenen, F., Laeven, H. and de Kruif, A. (2004) Transmission of classical swine fever. A Review. *Veterinary Quarterly* 26(4), 146–155.

Rojas, J.M., Sevilla, N. and Martín, V. (2021a) A new look at vaccine strategies against PPRV focused on adenoviral candidates. *Frontiers in Veterinary Science* 8, 729879.

Rojas, J.M., Martín, V. and Sevilla, N. (2021b) Vaccination as a strategy to prevent bluetongue virus vertical transmission. *Pathogens* 10(11), 1528.

Rolin, A.I., Berrang-Ford, L. and Kulkarni, M.A. (2013) The risk of Rift Valley fever virus introduction and establishment in the United States and European Union. *Emerging Microbes & Infections* 2(1), 18.

Romoser, W.S., Oviedo, M.N., Lerdthusnee, K., Patrican, L.A., Turell, M.J. *et al.* (2011) Rift Valley fever virus-infected mosquito ova and associated pathology: Possible implications for endemic maintenance. *Research and Reports in Tropical Medicine* 2, 121–127.

Rowlands, R.J., Michaud, V., Heath, L., Hutchings, G., Oura, C. *et al.* (2008) African swine fever virus isolate, Georgia, 2007. *Emerging Infectious Diseases* 14(12), 1870–1874.

Rüegg, S.R., Welby, S., Yassin, H., Van der Stede, Y., Nafzger, R. *et al.* (2018) Optimising cost-effectiveness of freedom from disease surveillance-bluetongue virus serotype 8 as an example. *Preventive Veterinary Medicine* 160, 145–154.

Rushton, J. and Lyons, N. (2015) Economic impact of bluetongue: A review of the effects on production. *Veterinaria Italiana* 51(4), 401–406.

Saegerman, C., Bonnet, S., Bouhsira, E., De Regge, N., Fite, J. *et al.* (2021) An expert opinion assessment of blood-feeding arthropods based on their capacity to transmit African swine fever virus in metropolitan France. *Transboundary and Emerging Diseases* 68(3), 1190–1204.

Sanson, R.L., Dubé, C., Cork, S.C., Frederickson, R. and Morley, C. (2014) Simulation modelling of a hypothetical introduction of foot-and-mouth disease into Alberta. *Preventive Veterinary Medicine* 114(3–4), 151–163.

Sasaki, Y., Alvarez, J., Sekiguchi, S., Sueyoshi, M., Otake, S. *et al.* (2016) Epidemiological factors associated to spread of porcine epidemic diarrhea in Japan. *Preventive Veterinary Medicine* 123, 161–167.

Sauter-Louis, C., Schulz, K., Richter, M., Staubach, C., Mettenleiter, T.C. *et al.* (2022) African swine fever: Why the situation in Germany is not comparable to that in the Czech Republic or Belgium. *Transboundary and Emerging Diseases* 69(4), 2201–2208.

Schley, D., Burgin, L. and Gloster, J. (2009) Predicting infection risk of airborne foot-and-mouth disease. *Journal of the Royal Society, Interface* 6(34), 455–462.

Schley, D., Whittle, S., Taylor, M. and Kiss, I.Z. (2012) Models to capture the potential for disease transmission in domestic sheep flocks. *Preventive Veterinary Medicine* 106(2), 174–184.

Schulz, C., Fast, C., Wernery, U., Kinne, J., Joseph, S. *et al.* (2019) Camelids and cattle are dead-end hosts for peste-des-petits-ruminants virus. *Viruses* 11(12), 1133.

SCoPAFF (2018). Available at: https://food.ec.europa.eu/document/download/72ea48ad-a38e-400d-82de-381c7faaf7a8_en?filename=reg-com_ahw_20181211_pdpf_bul.pdf (accessed 15 December 2024).

Sellers, R. and Gloster, J. (2008) Foot-and-mouth disease: A review of intranasal infection of cattle, sheep and pigs. *Veterinary Journal* 177(2), 159–168.

Sellers, R.F. and Parker, J. (1969) Airborne excretion of foot-and-mouth disease virus. *Epidemiology and Infection* 67(4), 671–677.

Sellman, S., Tildesley, M.J., Burdett, C.L., Miller, R.S., Hallman, C. *et al.* (2020) Realistic assumptions about spatial locations and clustering of premises matter for models of foot-and-mouth disease spread in the United States. *PLoS Computational Biology* 16(2), e1007641.

Sen, A., Saravanan, P., Balamurugan, V., Rajak, K.K., Sudhakar, S.B. *et al.* (2010) Vaccines against peste des petits ruminants virus. *Expert Review of Vaccines* 9(7), 785–796.

Seo, I.H., Lee, I.B., Hong, S.W., Noh, H.S. and Park, J.H. (2015) Web-based forecasting system for the airborne spread of livestock infectious disease using computational fluid dynamics. *Biosystems Engineering* 129, 169–184.

Seufi, A.M. and Galal, F.H. (2010) Role of *Culex and Anopheles* mosquito species as potential vectors of Rift Valley fever virus in Sudan outbreak, 2007. *BMC Infectious Diseases* 10(1), 1–8.

Shumilova, I., Krotova, A., Nesterov, A., Byadovskaya, O., van Schalkwyk, A. *et al.* (2022) Overwintering of recombinant lumpy skin disease virus in northern latitudes, Russia. *Transboundary and Emerging Diseases* 69(5), e3239–e3243.

Shurson, G.C., Palowski, A., van de Ligt, J.L.G., Schroeder, D.C., Balestreri, C. *et al.* (2022) New perspectives for evaluating relative risks of African swine fever virus contamination in global feed ingredient supply chains. *Transboundary and Emerging Diseases* 69(1), 31–56.

Simancas-Racines, A., Cadena-Ullauri, S., Guevara-Ramírez, P., Zambrano, A.K. and Simancas-Racines, D. (2023) Avian influenza: Strategies to manage an outbreak. *Pathogens* 12(4), 610.

Simons, R.R.L., Croft, S., Rees, E., Tearne, O., Arnold, M.E. *et al.* (2019) Using species distribution models to predict potential hot-spots for Rift Valley fever establishment in the United Kingdom. *PLoS One* 14(12), e0225250.

Sindato, C., Pfeiffer, D.U., Karimuribo, E.D., Mboera, L.E., Rweyemamu, M.M. *et al.* (2015) A spatial analysis of Rift Valley fever virus seropositivity in domestic ruminants in Tanzania. *PLoS One* 10(7), e0131873.

Sindato, C., Stevens, K.B., Karimuribo, E.D., Mboera, L.E.G., Paweska, J.T. *et al.* (2016) Spatial heterogeneity of habitat suitability for Rift Valley fever occurrence in Tanzania: An ecological niche modelling approach. *PLoS Neglected Tropical Diseases* 10(9), e0005002.

Singh, R.K., Balamurugan, V., Bhanuprakash, V., Sen, A., Saravanan, P. *et al.* (2009) Possible control and eradication of peste des petits ruminants from India: Technical aspects. *Veterinaria Italiana* 45(3), 449–462.

Song, H., Guo, L., Jin, Z. and Liu, S. (2022) Modelling and stability analysis of ASFV with swill and the virus in the environment. *Mathematical Biosciences and Engineering* 19(12), 13028–13049.

Souley Kouato, B., De Clercq, K., Abatih, E., Dal Pozzo, F., King, D.P. *et al.* (2018) Review of epidemiological risk models for foot-and-mouth disease: Implications for prevention strategies with a focus on Africa. *PLoS One* 13(12), e0208296.

Sprygin, A., Artyuchova, E., Babin, Y.U., Prutnikov, P., Kostrova, E. *et al.* (2018) Epidemiological characterization of lumpy skin disease outbreaks in Russia in 2016. *Transboundary and Emerging Diseases* 65(6), 1514–1521.

Ssematimba, A., Hagenaars, T.J., de Wit, J.J., Ruiterkamp, F., Fabri, T.H. *et al.* (2013) Avian influenza transmission risks: Analysis of biosecurity measures and contact structure in Dutch poultry farming. *Preventive Veterinary Medicine* 109(1–2), 106–115.

Stegeman, A., Elbers, A.R.W., Smak, J. and de Jong, M.C.M. (1999) Quantification of the transmission of classical swine fever virus between herds during the 1997–1998 epidemic in The Netherlands. *Preventive Veterinary Medicine* 42(3–4), 219–234.

Steinrigl, A., Fernández, S.R., Stoiber, F., Pikalo, J., Sattler, T. *et al.* (2015) First detection, clinical presentation and phylogenetic characterization of porcine epidemic diarrhea virus in Austria. *BMC Veterinary Research* 11(1), 310.

Stenfeldt, C., Pacheco, J.M., Singanallur, N.B., de Carvalho Ferreira, H.C., Vosloo, W. *et al.* (2015) Clinical and virological dynamics of a serotype O 2010 South East Asia lineage foot-and-mouth disease virus in sheep using natural and simulated natural inoculation and exposure systems. *Veterinary Microbiology* 178(1–2), 50–60.

Stenfeldt, C., Pacheco, J.M., Singanallur, N.B., Vosloo, W., Rodriguez, L.L. *et al.* (2019) Virulence beneath the fleece: A tale of foot-and-mouth disease virus pathogenesis in sheep. *PLoS One* 14(12), e0227061.

Stoian, A.M., Zimmerman, J., Ji, J., Hefley, T.J., Dee, S. *et al.* (2019) Half-life of African swine fever virus in shipped feed. *Emerging Infectious Diseases* 25(12), 2261–2263.

Sumaye, R., Jansen, F., Berkvens, D., De Baets, B., Geubels, E. *et al.* (2019) Rift Valley fever: An open-source transmission dynamics simulation model. *PLoS One* 14(1), e0209929.

Sun, R.Q., Cai, R.J., Chen, Y.Q., Liang, P.S., Chen, D.K. *et al.* (2012) Outbreak of porcine epidemic diarrhea in suckling piglets, China. *Emerging Infectious Diseases* 18(1), 161–163. DOI: 10.3201/eid1801.111259.

Su, S., Bi, Y., Wong, G., Gray, G.C., Gao, G.F. *et al.* (2015) Epidemiology, evolution, and recent outbreaks of avian influenza virus in China. *Journal of Virology* 89(17), 8671–8676.

Swayne, D.E., Suarez, D.L. and Sims, L. (2020) Influenza. In: Swayne, E.D., Boulianne, M., Logue, C.M., McDougald, L.R., Nair, V. *et al.* (eds) *Diseases of Poultry*, 4th edn. Wiley, Ames, pp. 210–256.

Tatar, N. and Alkan, F. (1999) Serological and virological investigation of rinderpest and pest of small ruminants virus in sheep and goats. *Etlik Veteriner Mikrobiyoloji Dergisi* 10, 35–60.

Tatar, N., Erturk, A., Kabakli, O., Akkoca, N., Incoglu, S. *et al.* (2002) Investigation of peste des petits ruminants prevalence in Turkey. *Etlik Veteriner Mikrobiyoloji Dergisi* 13, 15–31.

Tennant, W.S., Cardinale, E., Cêtre-Sossah, C., Moutroifi, Y., Le Godais, G. *et al.* (2021) Modelling the persistence and control of Rift Valley fever virus in a spatially heterogeneous landscape. *Nature Communications* 12(1), 5593.

Theuns, S., Conceição-Neto, N., Christiaens, I., Zeller, M., Desmarets, L.M.B. *et al.* (2015) Complete genome sequence of a porcine epidemic diarrhea virus from a novel outbreak in Belgium, January 2015. *Genome Announcements* 3(3), e00506–15.

Thornley, J.H.M. and France, J. (2009) Modelling foot and mouth disease. *Preventive Veterinary Medicine* 89(3–4), 139–154.

Tildesley, M.J. and Keeling, M.J. (2008) Modelling foot-and-mouth disease: A comparison between the UK and Denmark. *Preventive Veterinary Medicine* 85(1–2), 107–124.

Torres-Vélez, F. and Brown, C. (2004) Emerging infections in animals – potential new zoonoses? *Clinics in Laboratory Medicine* 24(3), 825–838.

Toyomaki, H., Sekiguchi, S., Sasaki, Y., Sueyoshi, M. and Makita, K. (2018) Factors associated with farm-level infection of porcine epidemic diarrhea during the early phase of the epidemic in Japan in 2013 and 2014. *Preventive Veterinary Medicine* 150, 77–85.

Tran, A., Trevennec, C., Lutwama, J., Sserugga, J., Gély, M. *et al.* (2016) Development and assessment of a geographic knowledge-based model for mapping suitable areas for Rift Valley fever transmission in Eastern Africa. *PLoS Neglected Tropical Diseases* 10(9), e0004999.

Trostle, P., Corzo, C.A., Reich, B.J. and Machado, G. (2022) A discrete-time survival model for porcine epidemic diarrhoea virus. *Transboundary and Emerging Diseases* 69(6), 3693–3703.

Tsao, K., Sellman, S., Beck-Johnson, L.M., Murrieta, D.J., Hallman, C. *et al.* (2020) Effects of regional differences and demography in modelling foot-and-mouth disease in cattle at the national scale. *Interface Focus* 10(1), 20190054.

Tukana, A. and Gummow, B. (2017) Dairy farm demographics and management factors that played a role in the re-emergence of brucellosis on dairy cattle farms in Fiji. *Tropical Animal Health and Production* 49(6), 1171–1178.

Tumwiine, J. and Robert, G. (2017) A mathematical model for treatment of bovine brucellosis in cattle population. *Journal of Mathematical Modeling* 5(2), 137–152.

Tun, H.M., Cai, Z. and Khafipour, E. (2016) Monitoring survivability and infectivity of porcine epidemic diarrhea virus (PEDv) in the infected on-farm earthen manure storages (EMS). *Frontiers in Microbiology* 7, 265.

Tuppurainen, E., Dietze, K., Wolff, J., Bergmann, H., Beltran-Alcrudo, D. *et al.* (2021) Vaccines and vaccination against lumpy skin disease. *Vaccines* 9(10), 1136.

Tuppurainen, E.S.M., Venter, E.H., Shisler, J.L., Gari, G., Mekonnen, G.A. *et al.* (2017) Review: Capripoxvirus diseases: Current status and opportunities for control. *Transboundary and Emerging Diseases* 64(3), 729–745.

Turner, J., Bowers, R.G. and Baylis, M. (2012) Modelling bluetongue virus transmission between farms using animal and vector movements. *Scientific Reports* 2, 319.

Valarcher, J.-F., Leforban, Y., Rweyemamu, M., Roeder, P.L., Gerbier, G. *et al.* (2008) Incursions of foot-and-mouth disease virus into Europe between 1985 and 2006. *Transboundary and Emerging Diseases* 55(1), 14–34.

van der Sluijs, M., Timmermans, M., Moulin, V., Noordegraaf, C.V., Vrijenhoek, M. *et al.* (2011) Transplacental transmission of bluetongue virus serotype 8 in ewes in early and mid gestation. *Veterinary Microbiology* 149(1–2), 113–125.

van der Sluijs, M.T.W., Schroer-Joosten, D.P., Fid-Fourkour, A., Vrijenhoek, M.P., Debyser, I. *et al.* (2013) Transplacental transmission of bluetongue virus serotype 1 and serotype 8 in sheep: Virological and pathological findings. *PLoS One* 8(12), e81429.

Van Oirschot, J. (1999) Classical swine fever. In: Straw, B.E., D'Allaire, S., Mengeling, W. and Taylor, D. (eds) *Diseases of Swine*, 8th edn. Ames, Iowa State University Press, pp. 159–172.

Vandenbussche, F., Mathijs, E., Philips, W., Saduakassova, M., De Leeuw, I. *et al.* (2022) Recombinant LSDV strains in Asia: Vaccine spillover or natural emergence? *Viruses* 14(7), 1429.

VanderWaal, K., Perez, A., Torremorell, M., Morrison, R.M. and Craft, M. (2018) Role of animal movement and indirect contact among farms in transmission of porcine epidemic diarrhea virus. *Epidemics* 24, 67–75.

Vergne, T., Gubbins, S., Guinat, C., Bauzile, B., Delpont, M. *et al.* (2021) Inferring within-flock transmission dynamics of highly pathogenic avian influenza H5N8 virus in France, 2020. *Transboundary and Emerging Diseases* 68(6), 3151–3155.

Wahlgren, J. (2011) Influenza A viruses: An ecology review. *Infection Ecology & Epidemiology* 1.

Wang, X., Sun, H. and Yang, J. (2021) Temporal-spatial analysis of an age-space structured foot-and-mouth disease model with Dirichlet boundary condition. *Chaos* 31(5), 053120.

Wang, Z., Bie, P., Cheng, J., Wu, Q. and Lu, L. (2015) *In vitro* evaluation of six chemical agents on smooth *Brucella melitensis* strain. *Annals of Clinical Microbiology and Antimicrobials* 14, 16.

Ward, M.P., Garner, M.G. and Cowled, B.D. (2015) Modelling foot-and-mouth disease transmission in a wild pig–domestic cattle ecosystem. *Australian Veterinary Journal* 93(1–2), 4–12.

Wareth, G., Abdeen, A., Ali, H., Bardenstein, S., Blasco, J.-M. *et al.* (2019) *Brucellosis in the Mediterranean Countries: History, Prevalence, Distribution, Current Situation and Attempts at Surveillance and Control.* World Organisation for Animal Health, Paris.

Weesendorp, E., Landman, W.J.M., Stegeman, A. and Loeffen, W.L.A. (2008) Detection and quantification of classical swine fever virus in air samples originating from infected pigs and experimentally produced aerosols. *Veterinary Microbiology* 127(1–2), 50–62.

Weesendorp, E., Stegeman, A. and Loeffen, W.L.A. (2009) Quantification of classical swine fever virus in aerosols originating from pigs infected with strains of high, moderate or low virulence. *Veterinary Microbiology* 135(3–4), 222–230.

Weesendorp, E., Backer, J. and Loeffen, W. (2014) Quantification of different classical swine fever virus transmission routes within a single compartment. *Veterinary Microbiology* 174(3–4), 353–361.

Whittle, L., Chapman, R. and Williamson, A.L. (2023) Lumpy skin disease – an emerging cattle disease in Europe and Asia. *Vaccines* 11(3), 578.

Wilson, A. and Mellor, P. (2008) Bluetongue in Europe: Vectors, epidemiology and climate change. *Parasitology Research* 103(S1), S69–77.

Wilson, A.J. and Mellor, P.S. (2009) Bluetongue in Europe: Past, present and future. *Philosophical Transactions of the Royal Society of London Series B, Biological Sciences* 364(1530), 2669–2681.

World Organisation for Animal Health (WOAH) (2024) First detection of peste des petits ruminants (PPR) in Greece and Romania. Available at: www.woah.org/en/first-detection-of-peste-des-petits-ruminants-ppr-in-greece-and-romania/ (accessed 11 December 2024).

Xiao, Y., Beier, J.C., Cantrell, R.S., Cosner, C., DeAngelis, D.L. *et al.* (2015) Modelling the effects of seasonality and socioeconomic impact on the transmission of Rift Valley fever virus. *PLoS Neglected Tropical Diseases* 9(1), e3388.

Yadav, S., Delgado, A.H., Hagerman, A.D., Bertram, M.R., Moreno-Torres, K.I. *et al.* (2022) Epidemiologic and economic considerations regarding persistently infected cattle during vaccinate-to-live strategies for control of foot-and-mouth disease in FMD-free regions. *Frontiers in Veterinary Science* 9, 1026592.

Yamada, M., Masujin, K., Kameyama, K.-I., Yamazoe, R., Kubo, T. *et al.* (2021) Experimental infection of pigs with different doses of the African swine fever virus Armenia 07 strain by intramuscular injection and direct contact. *Journal of Veterinary Medical Science* 82(12), 1835–1845.

Yamaji, R., Saad, M.D., Davis, C.T., Swayne, D.E., Wang, D. *et al.* (2020) Pandemic potential of highly pathogenic avian influenza clade 2.3.4.4 A(H5) viruses. *Reviews in Medical Virology* 30(3), e2099.

Zahur, A., Ullah, A., Irshad, H., Farooq, M., Hussain, M. *et al.* (2009) Epidemiological investigations of a peste des petits ruminants (PPR) outbreak in Afghan sheep in Pakistan. *Pakistan Veterinary Journal* 29(4), 174–178.

Zentkovich, M.M., Nelson, S.W., Stull, J.W., Nolting, J.M. and Bowman, A.S. (2016) Inactivation of porcine epidemic diarrhea virus using heated water. *Veterinary and Animal Science* 1–2, 1–3.

Zewdie, G., Derese, G., Getachew, B., Belay, H. and Akalu, M. (2021) Review of sheep and goat pox disease: Current updates on epidemiology, diagnosis, prevention and control measures in Ethiopia. *Animal Diseases* 1(1), 28.

Zhang, W., Zhang, J., Wu, Y.P. and Li, L. (2019) Dynamical analysis of the SEIB model for *Brucellosis* transmission to the dairy cows with immunological threshold. *Complexity* 2019(1), 6526589.

Zhang, Y., Chen, Y., Zhou, J., Wang, X., Ma, L. *et al.* (2022) Porcine epidemic diarrhea virus: An updated overview of virus epidemiology, virulence variation patterns and virus–host interactions. *Viruses* 14(11), 2434.

Zhao, J., Zhao, Y. and Zhang, G. (2023) Key aspects of coronavirus avian infectious bronchitis virus. *Pathogens* 12(5), 698.

Zientara, S., Sailleau, C., Viarouge, C., Höper, D., Beer, M. *et al.* (2014) Novel bluetongue virus in goats, Corsica, France, 2014. *Emerging Infectious Diseases* 20(12), 2123–2125.

Index